Sameera Kapila

INCLUSIVE DESIGN COMMUNITIES

PRAISE FOR INCLUSIVE DESIGN COMMUNITIES

This book is a gift. In these pages, Sameera has sketched out a compelling vision for the design industry's future. What's more, she's given us a map that—if we put in the effort to use it—leads toward a better, more inclusive version of that future.
— **Ethan Marcotte, author of *Responsive Web Design* and *You Deserve a Tech Union***

I have a deep and genuine love for those who can take a complex topic and make it actionable. As the need for more inclusive design communities grows daily, Sameera's straightforward, intentional approach meets the call. I love this book and I'm a better person for reading it.
— **Ryan Rumsey, author of *Business Thinking for Designers* and founder of Second Wave Dive and CDO School**

Whether you're experienced with or new to diversity, equity, and inclusion work, Inclusive Design Communities is an essential resource. Sameera's caring yet direct approach offers clear steps to evolve our design communities, proposing that we start the work within ourselves. Designers, technologists, human beings: Keep this on your bookshelf for clear-eyed reference and a welcome shift in perspective.
— **Raquel Breternitz, design leader and strategist**

Buy this book. The end.
— **Nathan Smith, design leader**

RELATED TITLES

Design for Safety
Eva PenzeyMoog

Sustainable Web Design
Tom Greenwood

Design for Cognitive Bias
David Dylan Thomas

Cross-Cultural Design
Senongo Akpem

You Deserve a Tech Union
Ethan Marcotte

Design for Real Life
Eric Meyer & Sara Wachter-Boettcher

Resilient Management
Lara Hogan

Accessibility for Everyone
Laura Kalbag

Copyright © 2025 Sameera Kapila
All rights reserved

Managing editor: Lisa Maria Marquis
Editors: Sally Kerrigan, Susan Bond, Kumari Pacheco, Caren Litherland
Technical editor: Mat Marquis

ISBN: 979-8-9913255-0-9

This title was originally published by A Book Apart and has been republished by the authors. Thank you to the A Book Apart team who first brought the book to life: Jeffrey Zeldman, Jason Santa Maria, and Katel LeDû, as well as Ron Bilodeau for producing the first edition.

10 9 8 7 6 5 4 3 2

TABLE OF CONTENTS

1 — *Introduction*

4 — CHAPTER 1
Understanding Inclusion

24 — CHAPTER 2
Expanding Mind and Action

36 — CHAPTER 3
Revisiting Design Education

55 — CHAPTER 4
Transforming Hiring Practices

82 — CHAPTER 5
Leading Inclusivity in the Workplace

107 — CHAPTER 6
Broadening the Community

130 — *Conclusion*

131 — *Acknowledgments*

133 — *Resources*

138 — *References*

146 — *Index*

*For Mom, Dad, Dee, and Ryan.
And for the ones who feel out of place (you're not!).*

FOREWORD

During my Women Talk Design workshops, I always ask participants about what draws them to public speaking. I hear many different answers, but nothing stopped me in my tracks the way one woman recently responded:

"Uninterrupted talk time."

She wasn't being heard at work, so she was eager to speak at conferences, meetups—anywhere there would be an opportunity for her to talk about her work without being interrupted. And while her words weren't surprising, they still landed like a punch in the gut. They underscored the many obstacles people with marginalized identities face in the design community.

Brilliant and important perspectives aren't being heard because our classrooms, workspaces, conferences, and clubs aren't spaces that include everyone. They aren't spaces where systematically marginalized people are invited to speak or contribute their ideas, because they aren't supported to feel safe, listened to, or like they belong.

But, there's good news: everyone can do something to make the design industry more inclusive, especially with Sameera Kapila guiding us.

Creating a more inclusive design community can feel like a daunting task, but Sam offers us a practical place to start. True inclusion is nuanced, specific, and ongoing; there's no checklist or quick fix. To do this work, we must shift our mindsets, reflect on our roles, and take action. We may sometimes get it wrong, and when we do, we must keep learning and growing, so we can bring our best selves to our communities—our education systems, hiring processes, meetings, and gathering spaces. This book is a first step on that journey, offering a wealth of resources that can serve us long after the final pages.

This work matters. Your role in it matters. So let's get started.

—Danielle Barnes

INTRODUCTION

Think of the communities you belong to. You might belong through proximity: neighbors who live near one another or gardeners who share a love of growing vegetables, fruits, and herbs. You might belong to a learning cohort that graduated the same year or a cadre of industry professionals who get together to talk shop every few weeks. You might feel a sense of community with groups of friends or with your immediate or extended family. Though there are bound to be differences among you—and those differences do matter—there is also a commonality that brings everyone together.

What about the design community? It's an ecosystem of individual and overlapping identities teeming with nuance and complexity. People in the design industry may identify as students, educators, job seekers, individual contributors, hiring managers, human resource partners, speakers, writers, mentors, award winners, judges, facilitators, retirees, conference organizers, club members, organizers, and so on.

Yet, for an industry that's supposed to be user-focused and forward-thinking, the design community isn't easily accessible or welcoming to all. Consider some of the obstacles that people new to the design industry face:

- **Access to education and training:** Whether you come from a traditional education route or a career change, there's often a steep and expensive learning curve. Designers pay to attend bootcamps, workshops, and conferences; students spend money on design software and hardware, classes, textbooks, and fonts. The design industry expects junior talent to have access to all of this.
- **Resistance to new perspectives:** Design conversations parallel art conversations in many ways, chock-full of opinions that have bounced between notable practitioners, institutions, and programs for decades. There is a clear preference for design that maintains the status quo. This creates a narrow view of design and inhibits the introduction and acceptance of new ideas.

- **Cultural gatekeeping:** Managing *who* is allowed to be *where* in the design space causes arguments over what titles designers can hold and what tools they should use. Becoming part of the industry often means fitting in—adapting to stereotypes and suppressing our authentic selves—and, in turn, enforcing that fit on others.

Stereotypes about who designers can be also feel like an obstacle that permeates our subconscious. For example, every time South by Southwest Interactive attendees flood the city where I live, I hear about what a designer is *supposed* to look like: white, male, flannel shirt, fancy laptop bag, and clear-framed glasses. Sometimes, I get the feeling that those who aren't white, cishet men aren't welcome in tech, or that they're expected to be in marketing, far away from design or code. I cringe every time a designer says an app needs to be "easy enough" for a grandmother—or a mother, or any woman of a certain age—to use, as if women are inherently incapable of using technology.

While these are just a few examples, they illustrate how alienating this community can be. People may choose to leave design as a result, or they may change themselves to better fit in. But in an industry that plays such a big part in how people interact, these dynamics are detrimental. Designers have a responsibility to include everyone in their circles and in the things they create.

Coretta Scott King, civil rights activist and wife of Dr. Martin Luther King Jr., once said: "The greatness of a community is most accurately measured by the compassionate actions of its members." Design—and, more generally, tech—tries to be compassionate, but there's a lot of work to do before intentions translate into actions. When we examine the needs of a community, and how those needs intersect with other communities, we can begin to find pain points and learn how best to address them.

In the coming chapters, we'll look specifically at three spaces within the design industry where people have formative experiences: the classroom, the workplace, and the community. We'll talk about how we can reduce—and work toward eliminating—the ways in which design is exclusionary, as we build more diverse design communities.

We each have an opportunity to apply what we know of design theory, instructional design, empathy, and leadership to make this industry better. Use this book to guide you toward more thoughtful decisions and to empower your own circles within design and beyond. By the end of this book, reader, I hope you feel empowered to make our field—and our world—a bit more inclusive.

1 UNDERSTANDING INCLUSION

Racism. Sexism. Ableism. Ageism. Classism. Nationalism. Homophobia. These few words represent many of the issues underpinning the design community—and many other communities. They carry weight, history, consequences, pain, and discomfort. They can appear anywhere—in the classroom, at a meeting or a conference, in social settings, at a new or existing job, or online. They are historically and deliberately woven into every fiber of almost every community, and design is no exception.

An inclusive design community is one where folks can reckon with a problematic past and take honest actions to amplify marginalized voices. But we must do the work, which entails:

- Seeking out what we don't know and reflecting on it
- Questioning—and disrupting—the status quo
- Actively and continuously fighting against exclusionary practices
- Advocating for others

Just as our duty as designers is to communicate with users and consider their needs, so inclusion is a practice of communication and consideration. We can't limit our design to the ideal user. We have to consider *everyone* who may come across it—and that means being aware of how we (do or don't) make space for designers in the industry. If we ensure our industry reflects the populations around us, we secure a way of better serving them.

Of course, it can be hard to figure out where to start with something so large and systemic. Inclusion can be intimidating in its impact and terminology—marginalization, for instance, calls for an entirely unique approach with space for reflection and healing. Before we dig deeper, though, let's run through some often misunderstood concepts that intersect with inclusion.

IDENTITY

When we talk about inclusion—and, by extension, exclusion—we must be aware of the countless aspects of identity found in a person's gender, age, race, socioeconomic status, sexual orientation, health, and more. Our identities shape how we show up in the world, and how we're perceived and treated.

I have been a student, a public- and private-sector educator, a manager, an executive, and a designer—and there are assumptions people might make about me based on those roles, especially those that connote "power" or "leadership." I am also a short, cishet female, India-born, Dutch-Caribbean-raised immigrant, millennial, and third-culture kid. Some of these identities may be readily apparent; others, such as my relationships with mental health and religion, are less visible.

Identities may also show up differently in certain settings, moving between dominant and marginalized positions. This means that situations can affect how we self-identify in terms of age, mental health, and ability, to name a few. And further, rarely does one person's identity exactly summarize everyone else with the same identity. What does it mean to navigate the design industry with these various identities? What happens when they overlap with certain environments and with one another?

Intersectionality

Coined by legal scholar Kimberlé Crenshaw in her 1989 essay "Demarginalizing the Intersection of Race and Sex: A Black Feminist Critique of Antidiscrimination Doctrine, Feminist Theory and Antiracist Politics" (https://samk.app/idc/01-01/), the term *intersectionality* has to do with how identities overlap and change the dynamic:

> *Intersectionality is a lens through which you can see where power comes and collides, where it interlocks and intersects. It's not simply that there's a race problem here, a gender problem here, and a class or LBGTQ problem there.* (https://samk.app/idc/01-02/)

In most industries, people with marginalized identities experience big monetary hurdles. In the United States, women get paid less than men for the same work, with lower rates of pay for Black, Hispanic, and Native women, according to a study by Payscale across salaried occupations (https://samk.app/idc/01-03/). This pay gap increases with age; it takes longer for women to work their way up the career ladder.

But pay is only one way disparities exist among different identities, and the intersection of said identities can change the landscape drastically. The same Payscale study showed that Asian women who are individual contributors may be paid more than white women, but are less likely to be promoted to director or executive roles. And a study by University of Massachusetts sociology professor Michelle J. Budig found that men are often paid more when they become parents (https://samk.app/idc/01-04/).

While the Payscale study recognized other factors—such as possibly skewed numbers due to the coronavirus pandemic, which affected women more than men—it did not consider other genders. While perhaps not intentional, the study itself marginalized nonbinary and gender nonconforming people by not capturing their data.

Appropriating identity

Sometimes identity is taken or colonized for other purposes. For instance, an ancient symbol meaning "wellbeing" or "fortune" for Hindus, Buddhists, and Jains was appropriated by Nazis during World War II, forever associating the swastika with genocide and hatred (https://samk.app/idc/01-05/). This is an example of *cultural appropriation*, "the adoption or exploitation of another culture by a more dominant culture," as defined by author Ijeoma Oluo in her book So You Want to Talk about Race (https://samk.app/idc/01-06/).

You might be familiar with this term from criticism of celebrities like the Kardashians, who are infamous for wearing Black hairstyles and overtanning their skin for the monetary benefit to their family businesses. Appropriation—in this case, *blackfishing*, a term coined by Wanna Thompson—is practically synonymous with their family (https://samk.app/idc/01-07/).

In the film industry, conversations often revolve around who should be allowed play certain roles and represent specific races, disabilities, and genders. A notable example was when Scarlett Johansson was cast in a Western remake of *Ghost in the Shell*, originally a Japanese movie with Japanese cast members. Fans cited this as *whitewashing*, a term (all too familiar in the American film industry) for when white actors play non-white roles, in this case using makeup to appear more Asian (https://samk.app/idc/01-08/).

An example of appropriation I've seen again and again is the whitewashing and commercialization of yoga and ritual-related words such as "om" and "namaste." Whenever I see a shirt that says "nama-slay" or "nama-stay-in-bed," my stomach turns to knots. It's such an intentional misuse, a joke directed at my culture. It's no longer relaxing to hear these words in yoga classes, let alone see them in clothing stores or on Pinterest. In "How 'Namaste' Flew Away From Us," NPR's *Code Switch* producer Kumari Devarajan wrote:

> *Sporting "namaste" on a water bottle or tote bag lets people present an essence and a persona that they believe is a part of an "exotic" culture simply by ... buying a tote bag. [...] When*

white English speakers fold words from other languages into their lexicon, they're often seen as cultured and worldly (and funny!). But for people of color, it's a totally different game. For example, after President Trump enacted the travel ban, Putcha says, her family became "hypervigilant" about speaking the language they normally spoke at home in public "for fear that it would raise suspicion" about their immigration status. (https://samk.app/idc/01-09/)

At best, cultural appropriation is theft; at worst, it erases identity and history. In early 2022, countries like Mexico created extensive laws around appropriation, recognizing the effect these trends have on communities (https://samk.app/idc/01-10/).

Appropriation isn't always obvious, however, especially in the design world. We're often "inspired" by what we see around us, making it our own and discarding its original meaning along the way—especially easy to do when it's for profit. We'll spend time breaking down intent versus impact later on, but no matter how you slice it, appropriation causes harm.

Erasing history

Like clothes and hair, words get appropriated into new contexts. Identity is sewn into the very fabric of our language. American idioms I've heard my whole life often have a troubled past—like the term "grandfathered," which shows up in conversations about everything from phone and internet billing plans to leasing agreements and laws. It originated with the "grandfather clause," a group of statutes in the 1890s that granted voting rights to direct descendants of anyone who had been eligible to vote prior to emancipation—automatically registering white men while requiring newly enfranchised Black men to pass tests or pay poll taxes in order to vote (https://samk.app/idc/01-11/). Sayings and idioms evolve over time, moving further and further from their original meaning. But the undertones of those meanings remain.

The modification of words and meanings is one way that history can be erased and reframed—often at the expense of marginalized identities. The malicious omission of historical fact is another. In 2020, many white parents and politicians protested the inclusion of topics like slavery, racism, and sexuality in American academic curricula. Books about reproductive healthcare, sexually transmitted diseases, teen parenting, trans rights, domestic abuse, suicide, and depression were removed from many schools and public libraries as a result (https://samk.app/idc/01-12/, https://samk.app/idc/01-13/). Pretending these topics don't exist doesn't make problems go away—it just creates harm by removing resources that people need. If we don't do our research, we're in danger of accepting these intentional erasures, promoting a history riddled with holes.

Appropriation, denial, and erasure of identities may vary in scale, but they are all active forms of marginalization. Let's take a step back and discuss marginalization, particularly in the context of microaggressions, safety, and intent.

MARGINALIZATION

Marginalization describes a person or group that is historically or presently pushed into the sidelines by a dominant or majority group, institution, or system. Most frequently, this term appears in the context of race and gender, but it can also occur with socioeconomic status, mental or physical health, ability, and so much more.

I use *marginalized* instead of a term like *minority* because marginalization emphasizes the harmful action taken against a person (or a group of people), rather than labeling a group as if they're outsiders. Whether it's the wheelchair user who's forced to use an out-of-the-way emergency exit to access a conference venue, the nonbinary person alienated by another "Ladies and Gentlemen" greeting, or the woman of color who is robbed of the chance to lead a meeting because she's "so good at taking notes," marginalization is embarrassing and belittling at best and dehumanizing at worst, making people feel like they don't belong in the spaces where they exist.

Marginalization can happen consciously or unconsciously, but the impact is harmful either way. It can change how people participate in classrooms and workplaces. Nothing about it feels great, and everyone loses. Unfortunately, marginalization exists in some form in every community. It has repercussions not only for the communities that stand to lose their members' participation, but for the identities directly touched by it.

If you're marginalized at work, your job may feel much harder. You may feel you have to prove yourself, because societal stereotypes suggest you look, act, and be a certain way. Stereotypes are destructive, particularly when it comes to what leadership looks like. When we think of a CEO or manager, many people still default to the image of a white, able-bodied male, and words like "professional" and "authoritative" are usually code for that same idea. Suppose you're in design leadership and don't meet any of those stereotypical identities. The job may be more challenging because you're not what people expected; they may discount your ability and expect you to fail, take missteps, or not do things the "right" way.

When you are made to feel that your differences are a liability or distracting, it's natural to want to try to blend in—which brings us to the soul-crushing habit of assimilating to an inherently hostile environment. It's also normal to feel the doubt and dread of imposter syndrome, unfortunately a common experience for anyone who isn't a white man in the tech industry. It can be a burden to show up every day and put on a figurative, protective mask; to shrink or camouflage for survival instead of just being able to exist and participate in the space like everyone else.

I don't begrudge anyone for taking steps to keep themselves psychologically (or physically) safe. But rather than asking marginalized individuals to show up "whole" (as if they prefer to show up partially!), let's call the problem what it is: an unsafe environment perpetuated by a desire for power to stay with those who already have it.

Microaggressions

One of the most common ways people are marginalized is through *microaggressions*. Microaggressions appear in many forms and may fly under the radar for those unfamiliar with them. They can be verbal or nonverbal, intentional or unintentional. They can show up as backhanded compliments, seemingly casual remarks about one's cultural background, charged stereotypes or bias framed as curiosity, denial of someone's identity, ignoring ideas and speaking over someone in a meeting, or seeking validation for previously held beliefs.

Microaggressions are tricky because they aren't always as blatantly obvious as other forms of "-isms." However, because of their persistent repetition, microaggressions can cause long-term, generational trauma. It's death by a thousand cuts. According to the Michigan State University Office for Inclusion and Intercultural Initiatives, the "accumulated impact of daily microaggressions cause real pain, anxiety, depression, self-doubt and [...] adverse health" (https://samk.app/idc/01-14/, PDF).

Countless forms of microaggressions exist. Here are a few examples of verbal microaggressions I've experienced:

- **"Where are you *from* from?"** I've been asked this question for as long as I've lived in the United States. Because of the country and ethnicity I was born into, people don't believe me when I tell them I grew up in a Dutch country in the Caribbean. Even though my Dutch is better than my Hindi and I've spent more time where I grew up than in India, people find it hard to believe that I define myself as Caribbean-raised. If you hail from a place where many people look like you, it may not seem like that big of a deal to ask someone "where they're *from* from." In reality, though, it can be highly offensive. By asking this question, you're revealing your own biases—specifically *confirmation bias*, where you're hoping someone will answer the question the way you expect them to answer, rather than providing an unfamiliar truth.
- **"Wow, you do that so well."** People are often amazed that I speak English when I didn't grow up in the US or Europe, or that my accent changes based on the languages or dialects I use. When you tell a coworker that they "speak English well," or that you're impressed they're able to do a task while using a wheelchair or crutches, it comes across as a back-

handed compliment—it implies you didn't think they were capable of it.
- **"Your name is hard to pronounce."** My sister and I were given the names *Deepina* and *Sameera* for several reasons: we were named after certain people, yes, but my parents also knew these names would be easier to shorten and westernize to *Dee* and *Sam*. To *westernize* is to adjust or adapt one's given name to make it easier for Europeans and Americans to pronounce. While more and more people are reclaiming their given names (https://samk.app/idc/01-15/), many immigrant families want to avoid having their children be "othered" or treated differently; Western educators often mispronounce non-Western names, or give out nicknames, not even bothering to learn students' given names. The same (and worse) happens outside the classroom, in workplaces and in civic life. Consider, for example, a Texas politician's suggestion that people of Asian descent should "adopt a name that [Americans] could deal with" (https://samk.app/idc/01-16/). This attitude puts the burden on the individual to change, rather than on everyone else to respect the individual's identity. All it takes is a simple, "I'd love to learn how to pronounce your name correctly," or "What name do you go by?"
- **"You're diverse."** I've seen this in the recent push for schools and companies to recruit people with marginalized identities into predominantly white spaces. A group of people—such as a team or student body—can be *diverse*, but an individual cannot. Referring to an individual person as *diverse*, explicitly or implicitly, is a way of othering them for their difference.

While microaggressions may seem trivial, they can cause the recipient significant pain. Multiple microaggressions can be enough to make someone feel unwelcome and unwanted in a space, make them doubt themselves, and, eventually, make them leave their job, industry, or community. This happens in the tech and design industries at an alarming rate.

Safety

When employees feel like they don't belong in a space, they might embody a state of constant vigilance. *Is it safe to be here? What if someone approaches me? Where are the exits if I need to leave suddenly? Do others think I'm a threat? Should I try to blend in?* Being from a marginalized community can make these questions especially loud, taking away one's ability to be fully present in an office, classroom, or conference space. In the Medium publication *AfroSapiophile*, writer Nini Mappo described this sense of constant evaluation:

> *To always be on high alert, scanning your surroundings and beyond, looking over your shoulder, is to nurse a relentless din of 'I may not be safe. I am not safe. I can't just relax.'* (https://samk.app/idc/01-17/).

People with marginalized identities might experience situations that trigger a "fight, flight, freeze, or fawn" response (https://samk.app/idc/01-18/). Stress activates the body's sympathetic nervous system to determine how danger should be dealt with—by confronting the scenario (fight), escaping it (flight), becoming unable to move or act (freeze), or trying to agree with or please the aggressor to avoid conflict (fawn). The same situation can elicit a stress response in bystanders as well: to defend the individual (fight), leave the situation entirely (flight), do nothing (freeze), or minimize the individual's resulting pain (fawn). All of these responses come with a fear of retaliation.

Some marginalized people may minimize or affirm their identity as a form of protection. Designer Timothy Bardlavens described his own approach to identity in the workplace in a Medium article called "Navigating Whiteness":

> *Corporate life taught me how to navigate whiteness by leveraging my jokey demeanor and my gayness to be perceived as less threatening as a Black man. It taught me to lean on being gay to befriend white women and turn them into "fag hags" because it allows us to communicate more effectively, all*

while understanding I could never confide in them or get angry around them. (https://samk.app/idc/01-19/)

Code-switching is a term that describes what happens when marginalized people adjust their language around different groups of people to better protect themselves:

> Research suggests that code-switching often occurs in spaces where negative stereotypes of black people run counter to what are considered "appropriate" behaviors and norms for a specific environment. For example, research conducted in schools suggests that black students selectively code-switch between standard English in the classroom and African-American Vernacular English (AAVE) with their peers, which elevates their social standing with each intended audience. We also see examples of guidelines encouraging black people to code-switch to survive police interactions, such as "acting polite and respectful when stopped" and "avoiding running even if you are afraid." (https://samk.app/idc/01-20/)

People with dominant identities must broaden their understanding of these dynamics. Coworkers, friends, even family members may be grappling with safety and a sense of belonging; they may use their survival instincts on a daily basis.

If you are from a marginalized community and are trying to navigate safety and belonging in your workplace or classroom, know this:

- **You deserve to be treated fairly.** You have every right to be in your workplace—or in any space you operate in—as much as anyone else. You have the right to speak up when you see issues and receive the same support and access as any other team member.
- **It's okay to share only what you want to share.** You don't have to answer any questions about any part of your identity if you don't want to. Set boundaries so that you're not responsible for educating others.
- **Assimilation is not required.** If you feel safe, you can choose to own your experience rather than trying to hide it or fit

into some box of what others think a designer should be. Your unique experiences will add to the design conversation and encourage new ways of thinking.
- **Document everything.** Keep digital or physical copies of your work (especially if you're an immigrant, as many work visas require samples). If you experience situations that are not inclusive, document the events or conversation, even if it's after the experience has ended. If you ever need to escalate something, keeping a record of screenshots, emails, notes, etc., may help you make a case for change. Although this undeniably adds more emotional and administrative labor to your day-to-day, it can be one of the best ways to defend yourself in a difficult situation.

When we recognize the importance of safety in our environments and understand the dynamics of identity, we can start to see the potential impact on historically oppressed or targeted people. However, another layer to add to this is knowing the difference between the intent of our behaviors and the impact they ultimately have.

Intent and impact

Intent—in the context of inclusivity—is the idea that we're doing something we think will help others. *Impact* is our action's actual effect on someone, how it makes them feel. Unfortunately, intent and impact are frequently not the same. You may have the best intentions, but if your actions make others feel differently than you'd intended, there can be consequences.

For example, at the code school where I worked, we offered a diversity scholarship for students who identified with groups frequently marginalized in tech. The scholarship application was a separate, longer form that asked students about their experiences. However, we found that this additional application just added hurdles for the students we most wanted to help. Our intent was good, but the impact wasn't.

We learned from this mistake and changed our application process, offering students the opportunity to self-identify on the main application form instead. While it wasn't a perfect solution, more students were able to apply for and receive scholarships.

On She+ Geeks Out, Fatima Dainkeh wrote about the differences between intent and impact, and how to work through negative impact (https://samk.app/idc/01-21/). Dainkeh suggested those with good intentions should:

- remember they can't control the responses of others,
- apologize sincerely when they cause harm,
- educate themselves on structural "-isms," and
- forgive themselves for making mistakes.

Those who have been impacted should:
- ask for clarification before responding,
- recognize that they're in control of their reaction,
- find support in trusted people, and
- remember it's not their job to educate those who have harmed them.

We all benefit from prioritizing an action's impact over its intention, especially when the impact doesn't measure up to said intention.

ALLYSHIP

Allyship is about unconditionally supporting others, specifically those who are oppressed. Being an ally requires internal and external work, and none of it is easy. Calling oneself an ally doesn't necessarily mean you are one; it's a word someone else gifts you, not one to self-proclaim. A friend once told me, "I'm not a safe person just because I say so."

Designer Amélie Lamont offers some guidance in their resource "Guide to Allyship" (https://samk.app/idc/01-22/). According to the guide, to be an ally is to:

1. Take on the struggle as your own.
2. Transfer the benefits of your privilege to those who lack it.
3. Amplify voices of the oppressed before your own.

4. Acknowledge that even though you feel pain, the conversation is not about you.
5. Stand up, even when you feel scared.
6. Own your mistakes and de-center yourself.
7. Understand that your education is up to you and no one else.

If you want to be a good ally, you need to show your commitment through your actions. The label itself isn't productive, but the work you do is. You have to consider all the granular details when thinking about equality, equity, and liberation as a form of allyship.

Self-centering

It's hard to be a good ally to another person when we focus on ourselves. Part of being an inclusive member of any space is understanding that centering yourself in the conversation—or an area of conflict or confrontation—can do more harm than good.

For example, let's say you've misgendered an employee or coworker. Once you've caught yourself or been corrected, a reasonable impulse is to apologize, maybe even excessively. While apologies are okay, centering yourself by apologizing too much makes it more about you and less about the person who was misgendered. It becomes their burden to manage your feelings after your mistake.

Centering yourself is one form of *performative allyship*. Unlike authentic allyship, performative allyship is for display only. It's common in public statements issued by corporations trying to join the cultural conversation, such as when the National Football League (NFL) used social media to say that "Black Lives Matter" while actively working to make sure Colin Kaepernick wouldn't become a starting quarterback again within the NFL (https://samk.app/idc/01-23/).

I've seen designers create fonts using the names of murdered Black Americans to promote their design work—many of which were taken down after social media pushback. Performative allyship tends to look good on Instagram and other social media where people create or repost statements of support, but it doesn't directly affect change or amplify marginalized voices. Instead, it adds noise and frustration to the issues it's supposed to be helping.

Remember: it's not about you. If you're spending a lot of time talking about your actions as an ally, you might be centering yourself in the conversation rather than truly focusing on those who need your support. The whole point of being an ally is to understand the systems and help shift the balance, not to become a hero.

Equality, equity, and liberation

As we continue to cover inclusion and allyship, we need to understand a few key terms related to systems: *equality*, *equity*, and *liberation*. These terms are often used interchangeably, but differ in crucial ways. An illustration of people trying to watch a baseball game from behind a fence (**Fig 1.1**) can act as a metaphor to explain the nuances:

- *Equality* means everyone is given the same opportunities—in this case, boxes to stand on to see over the fence. Most people in the design industry stop at equality and think it's enough. But the notion of equality doesn't address any systemic hurdles or issues—even with the box, not everyone can see over the fence, and some people don't need the box to begin with.
- *Equity* is the idea that everyone is treated *fairly*, not just equally. It considers experiences and needs, highlighting the systemic variables causing injustice or unfairness in relation to our identities. In the fence metaphor, equity means everyone receives the number of boxes necessary to improve their own view.
- While equality and equity address dealing with injustice in their own ways, the third term, *liberation*, suggests the complete removal of injustices—in the case of this metaphor, eradicating the fence entirely.

Fig 1.1: Helping people of different heights see a ballgame over a fence can be a metaphor for equality, equity, and liberation. This illustration is a collaboration between the Center for Story-based Strategy (https://samk.app/idc/01-24/) and Interaction Institute for Social Change (https://samk.app/idc/01-25/).

The final panel of the illustration is a blank space for additional ideas—where the two collaborating institutions hope people will think about what step could come next in pursuing justice.

Privilege

Being an ally means you're committed to taking an honest, unflinching look at your privilege. You get to decide how to use your privilege to support people in more vulnerable situations and uplift marginalized voices.

The very word *privilege* tends to make people defensive or think narrowly—many associate privilege solely with having access to wealth. While that's part of it, privilege can also refer to a person's race, skin color, ability, education, location, gender, and much more. The concept dates back to as early as 1903, when W. E. B. Du Bois used it in *The Souls of Black Folk*. In 1988, Peggy McIntosh published a groundbreaking scholarly article on the topic, pushing the term further into the public eye (https://samk.app/idc/01-26/).

Financial privilege is one of the easier forms to identify because it's frequently tied to material things like an expensive car, brand-new shoes, or electronics—things we believe we've earned or deserve. Educator, speaker, and writer Marie Beecham frames it another way:

> *Some people have a hard time recognizing privilege, saying "I work hard. I don't get things handed to me." I understand that. Here's how I respond: privilege isn't bonus points for you and your team. It's unfair penalties the other team gets that you don't.*
>
> *Privilege isn't the presence of perks and benefits. It's the absence of obstacles and barriers. That's a lot harder to notice. If you have a hard time recognizing your privileges, focus on what you don't have to go through. Let that fuel your empathy and action.* (https://samk.app/idc/01-27/)

For example, every time I get coffee or tea at a shop nearby, I don't have to worry about whether I'll encounter stairs. They don't pose an obstacle for me to get to where I need to go. But for someone in a wheelchair or who uses crutches after an injury, stairs are physical obstacles that put them at a disadvantage if there isn't a ramp available. In this scenario, I have privilege; the coffee shop isn't as accessible to others as it is to me.

You might find yourself feeling some level of guilt because of the privileges you have. We must confront such guilt head-on and acknowledge that it exists, rather than letting it pull us back into self-absorption and hinder us from doing the critical work of self-improvement. It isn't helpful to feed the narrative that you can't help what you were born into. Instead, understand that guilt is a powerful motivator, telling us where we have opportunities to make things better. Use your privilege (or guilt) to elevate others who have been historically marginalized and oppressed. At work, you might intervene in a tense meeting or shine the light on others who get spoken over frequently, asking the interrupting party to let the other person complete their thought.

Designer Timothy Goodman tries to use his social media popularity—his privilege—to share professional opportunities with marginalized designers. For example, when an agency asked him to contribute to a campaign for "underrepresented" groups, he pointed out that such a campaign would do better if they chose to "[hire] artists who actually represent these groups" (**Fig 1.2**):

> *Your brand or agency's advocacy is not enough if you're not actually paying artists who represent the groups you're making social campaigns for. I'm still surprised when I have to write emails like this, and it's important to keep publicly talking about it. I'm not perfect, but more of us straight cis white guys need to stop taking this kinda work and educate our clients.* (https://samk.app/idc/01-28/)

This is an excellent example of how designers can apply their skill sets to be better allies.

As designers, we should always be looking for ways to directly support marginalized communities. Making posters and joining Slack channels can only go so far. Can we mentor a student whose educators aren't helping them succeed? Can we interrupt a meeting when a manager speaks over a frequently excluded employee? Can we stop insensitive jokes while they're happening? Can we suggest that designers with dominant identities step aside to allow those from marginalized commu-

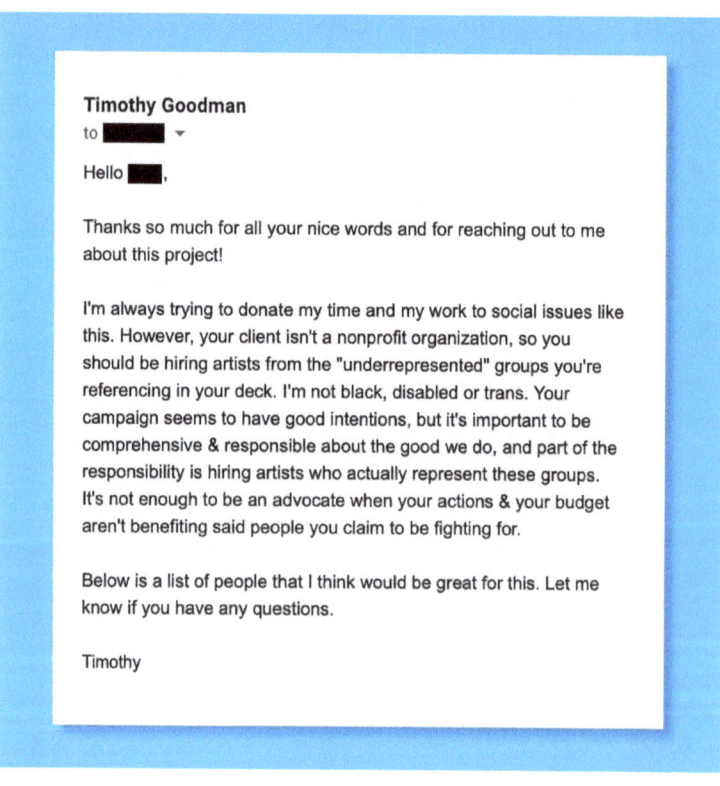

Fig 1.2: An email response Timothy Goodman sent to one of his prospective clients. In the response, Goodman points out that the client has the money to hire artists from the groups they're campaigning for, while adding that he himself does not belong to those groups. He also offers a list of people the client can reach out to instead.

nities to represent themselves in work that reflects their identities?

We all have privileges in certain areas and in certain contexts. We must be able to talk about them if we have any hope of leveling the playing field for everyone.

SO, WHAT NOW?

Inclusion means respecting the way people interpret their value within a community, such as the design industry. We should all have a seat at the table. We should all feel comfortable enough to bring our whole selves to the spaces we're in and to invite others to them, too. We should make sure everyone around us is represented and treated equitably. After all, opportunities are all around us.

As designers, we work a lot with systems: design systems, brand and identity systems, advertising campaigns, and more. So when we think of our community as a system, we put ourselves in the position to take more logical steps toward equity and liberation.

In the following chapters, we'll learn how identity, marginalization, and inclusion play into various design spaces such as educational institutions, workplaces, conferences, and more. But before we break into those spaces, we'll dive into more approaches for exacting change individually.

2 EXPANDING MIND AND ACTION

In her book *Why Are All the Black Kids Sitting Together in the Cafeteria?: And Other Conversations About Race*, Beverly Daniel Tatum wrote:

> I sometimes visualize the ongoing cycle of racism as a moving walkway at the airport. Active racist behavior is equivalent to walking fast on the conveyor belt. The person engaged in active racist behavior has identified with the ideology of white supremacy and is moving with it. Passive racist behavior is equivalent to standing still on the walkway. No overt effort is being made, but the conveyor belt moves the bystanders along to the same destination as those who are actively walking. Some of the bystanders may feel the motion of the conveyor belt, see the active racists ahead of them, and choose to turn around, unwilling to go in the same destination as the White supremacists. But unless they are walking actively in the opposite direction at a speed faster than the conveyor belt—unless they are actively antiracist—they will find themselves carried along with the others.

I'm invoking this quote not solely on the point of racism, but of all the "-isms." If we want to combat our community's "-isms"—racism, sexism, ageism, ableism, etc.—we can't just go with the flow. It's necessary to move in an inclusive direction with intention.

Being intentional about our growth can slowly but surely impact the people and communities around us. But trying to change external factors without inward reflection can do more harm than good. The work begins with growing our motivations, opening our minds, assessing what we know, and understanding that some of this work is uncomfortable but beneficial.

SHIFTING YOUR MINDSET

Before we try to change a community (or with luck, the world), we need to start with ourselves. Focusing on the work we can do in our thinking, and eventually in our actions, gives us control over what we can *actually* change—which is ourselves.

Prepare for growth

Taking on this sort of uneasy self-work requires a growth mindset, rather than a fixed mindset. I first learned these terms from Maria Popova's design blog, *Brain Pickings*, where she discusses the work of researcher, professor, and author Carol Dweck. Popova describes a *fixed mindset* as the belief that:

> Our character, intelligence, and creative ability are static givens which we can't change in any meaningful way, and success is the affirmation of that inherent intelligence, an assessment of how those givens measure up against an equally fixed standard; striving for success and avoiding failure at all costs become a way of maintaining the sense of being smart or skilled. (https://samk.app/idc/02-01/)

The fixed mindset focuses on the illusion of safety by avoiding failure and keeping its world small and controllable—which is not the way to invite diverse viewpoints. In any case, failure

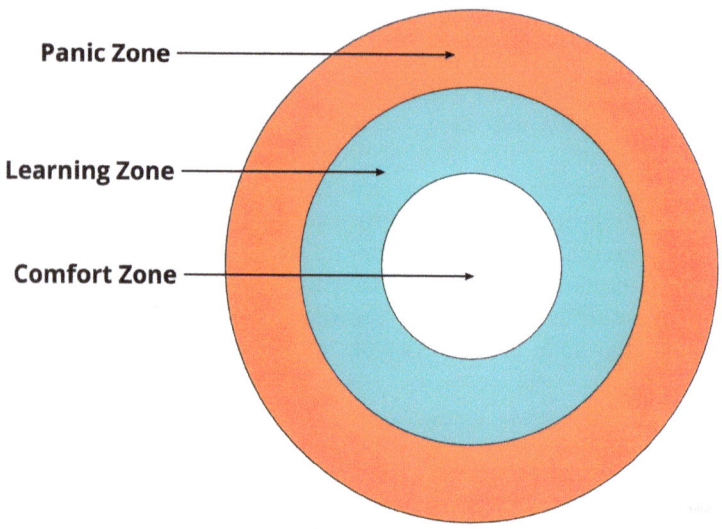

Fig 2.1: The Learning Zone Model suggests the best place to learn is beyond our comfort zones, but not so far beyond them that we panic.

usually starts as an internal emotion unseen by others. Fear keeps us in a fixed mindset, like the freeze response to trauma mentioned earlier. A fixed mindset also doesn't allow biases to dissolve.

A *growth mindset*, on the other hand, allows us to be more reflective and to consider that, though our biases may be flawed, they can be unlearned. Popova goes on to say that the growth mindset:

> *thrives on challenge and sees failure not as evidence of unintelligence but as a heartening springboard for growth and for stretching our existing abilities.*

A growth mindset creates space for us to ask questions, think critically, and check in with ourselves. It creates time to be intentional and gives us space to listen to others.

When I try to visualize a growth mindset, I employ a tool called the Learning Zone Model (https://samk.app/idc/02-02/). The idea behind the Learning Zone Model, a term coined by Lev Vygotsky, is that when we're learning, there is a *comfort zone* that feels safe to us, a *panic zone* that feels unfamiliar, and a space in between called the *learning zone* (**Fig 2.1**).

Visualizing these zones can help us check in with ourselves, become aware of when we feel uncomfortable, and find a way back into the learning zone where our most open-minded thinking happens. If the concepts are too easy and we're too comfortable, we're not growing; if they're too difficult, we panic. Finding a balance between the two is how we keep learning.

Reflect on your biases

We can think about bias in two ways: explicit biases and implicit biases. *Explicit biases* are conscious opinions. You can be biased toward hometown sports teams because you share a common location. You can be biased against someone who went to a rival school because of past competition. You can favor dogs over cats, or express bias for your dogs versus your neighbor's. You might be biased toward or against certain typefaces, user interface patterns, or software programs.

Implicit biases are subconscious, meaning we may be unaware we have them. In *Design for Cognitive Bias* (https://samk.app/idc/02-03/), David Dylan Thomas wrote that implicit bias is part of an autopilot decision-making process that comes from our life experiences. Implicit biases have consequences even when we don't know we're acting on them. For example, a coworker might always speak over you in meetings because of a subconscious belief about who holds authority in professional settings.

Project Implicit—a collaborative research project between Harvard University, the University of Virginia, and the University of Washington—assesses bias through an Implicit Association Test (IAT):

> The IAT measures the strength of associations between concepts (e.g., black people, gay people) and evaluations (e.g., good,

bad) or stereotypes (e.g., athletic, clumsy). The main idea is that making a response is easier when closely related items share the same response key. When doing an IAT you are asked to quickly sort words into categories that are on the left and right hand side of the computer screen by pressing the "e" key if the word belongs to the category on the left and the "i" key if the word belongs to the category on the right. (https://samk.app/idc/02-04/)

The way test-takers match words and categories can gauge how they make subconscious associations. It also illustrates how, in our day-to-day lives, those judgments may kick in before anything else:

While a single IAT is unlikely to be a good predictor of a single person's behavior at a single time point, across many people the IAT does predict behavior in areas such as discrimination in hiring and promotion, medical treatment, and decisions related to criminal justice. (https://samk.app/idc/02-05/)

Part of understanding bias is learning to notice it and reframe it. There are many ways we can try to become more aware of, and thus reduce, implicit bias:

- Taking a step back before making decisions
- Pausing in the middle of an action
- Listening to people who see things differently
- Interrogating our motivations and decisions
- Removing assumptions or judgment from communications
- Allowing people to be who they are

These are all ways to move toward changing our behavior for the better.

Own your mistakes

No one will get all of this right on the first try. The fact is, mistakes are a given. I've made countless mistakes, some of which you will learn about in this book. It might be helpful to remember the first time you made a mistake while designing or coding. You learned something from it, right? You might have beaten yourself up, but you didn't abandon your work. You kept going. You likely didn't make the same mistake again, or if you did, you recovered more easily.

Mistakes are part of the journey. Accepting them and being open to learning and progressing applies across all contexts. Author and educator bell hooks said:

> *In all cultural revolutions, there are periods of chaos and confusion, times when grave mistakes are made. [...] We cannot be easily discouraged. We cannot despair when there is conflict. Our solidarity must be affirmed by shared belief in a spirit of intellectual openness that celebrates diversity, welcomes dissent, and rejoices in collective dedication to the truth.* (https://samk.app/idc/02-06/)

When you understand that mistakes may happen, but you move forward anyway, that's when you know you're in the learning zone. If you don't aspire to work through the side effects of failure, you can end up creating more harm.

In the early days of the #MeToo movement, when more women started speaking out about workplace harassment, too many male managers feared being thrust into the public eye because of an unintentional misstep. Their response was to swear off mentoring or supporting female team members altogether. The *New York Times* cited two 2019 studies surveying almost nine thousand adults, and the results were eye-opening:

> *Almost half of male managers were uncomfortable engaging in one or more common work activities with women, such as working one on one or socializing. One in six male managers was uncomfortable mentoring a female colleague.* (https://samk.app/idc/02-07/)

This is a complex situation. Let's begin by asserting that, yes, it's a mistake—actually, a crime—to harass or assault colleagues, and to target people with less perceived power in professional circles. But men who refuse to support the women they mentor and manage are letting fear govern their reaction, centering their own feelings ("I might be called out and embarrassed!") over the reality that, without any support, female and nonbinary reports are placed at a real disadvantage.

This is also a good example of the disconnect between intent and impact. The intent is to avoid inappropriate situations in the workplace, but the impact is protecting managers while leaving employees unsupported. It also shifts responsibility and blame from the managers, who have more power, and places it on the reports, who have very little. Intent doesn't matter with such a harmful impact.

Reasonable mistakes happen to the best of us. We are imperfect people, and we're going to make mistakes when it comes to inclusion. What matters is what we do after we learn we've made a mistake. It's about owning it, apologizing for it, and correcting it moving forward.

TAKING ACTION

Now that we've shifted our mindset, it's crucial to turn our thinking into action. In *Teaching to Transgress: Education as the Practice of Freedom*, bell hooks wrote about how critical it is to go beyond making a verbal commitment to fighting injustice:

> I have encountered many folks who say they are committed to freedom and justice for all even though the way they live, the values and habits of being they institutionalize daily, in public and private ritual, help maintain the culture of domination, help create an unfree world. (https://samk.app/idc/02-06/)

In this section, we'll explore how to apply what we just learned to our daily actions. When those more intentional, inclusive actions become habits, we improve how we show up in our different spaces, making others feel seen, heard, and supported.

Watch your language

Language is a powerful tool. There are words that are hurtful and painful, and there are words that are supportive and inviting. Our communication can lead people to feel included or excluded—or it can be weaponized to maintain the status quo.

Consider some common examples of how language is used to include some groups while excluding others:

- Women in leadership positions are seen as "bossy" or "demanding," while the same qualities in men are described as "assertive" and "in charge."
- "Illegal immigrant" is a term often used by anti-immigration proponents to draw attention to what they see as a criminal action: not having certain documentation. However, immigrants are people, and *people* are not against the law. Furthermore, immigrants may lack documentation for many reasons, few of which are actually considered criminal. There is a stark difference between saying "illegal immigrant" and "undocumented immigrant."
- The media often describes a white shooter as a "lone wolf," while a shooter with darker skin is a "terrorist" or a "thug." In an article for *Quartz*, writer Youyou Zhou found the term "lone wolf" was used almost exclusively to refer to white shooters: "White people were rarely described as radicalized individuals. The word 'radical' was not mentioned once in the 11 of 27 incidents in which the killers were identified as white, but it appeared 33 times when non-white killers were involved (https://samk.app/idc/02-08/)."

These examples show both conscious and unconscious ways that words can warp our understanding of—or empathy for—others. Word choice can change public narratives, romanticizing the actions of white people while punishing people of color for the same—and often less dangerous—behaviors. This can influence everything from the way people are portrayed in the media to legal prosecution and sentencing. Time after time, regardless of the crime, the consequences for people of color are higher than for their white counterparts, even when their crimes are less severe (https://samk.app/idc/02-09/). The words we use need scrutiny.

Language is part of design, too. As designers, we leave our mark on what we are helping to communicate; it's our responsibility to be mindful of the words we use when we set type for print or screens. Are we using words for shock value or for our own pleasure? Are we checking with others to see if the language could be harmful? Are our biases coming into play? How can we update the words we use to invite more viewers and users? How might we consider their identities and the situations they're in when we're designing? It behooves us to move past throwing words into our work without examining their implications.

Listen actively

One of the most effective ways to be a good leader is to practice active listening. Active listening makes people feel heard, which is validating for people experiencing marginalization. It also promotes psychological safety, empowering people to present their own ideas without fear of punishment or derision.

One of the key elements of active listening is creating space for coworkers to express themselves. They can do that when they feel they are able to trust and be heard by others. The Center for Creative Leadership (CCL) defines active listening as consisting of six steps (https://samk.app/idc/02-10/):

1. **Pay attention.** Focus on what is being said and wait before responding. Also, be aware of the body language and mindset you are listening with.
2. **Withhold judgment.** Be open-minded and mindful of the speaker's opinions, even if you disagree.
3. **Reflect.** Repeat what the speaker is saying to demonstrate your understanding and facilitate their thinking.
4. **Clarify.** Ask open-ended questions that clear up ambiguity and support understanding.
5. **Summarize.** Restate the conversation to make sure the speaker has been fully heard.
6. **Share.** Offer your own ideas and responses.

Active listening can help build trust, a mark of a good ally and leader. Taking the time to listen can create more space for honest and meaningful conversations.

Intervene

People who are marginalized need others they can rely on to stand up for them, particularly those with more privilege or power. Standing up doesn't mean donating to a fundraiser or posting online (although those actions may be helpful in other ways); it means supporting others through active intervention in situations that call for it.

If you notice a coworker is interrupting your teammate constantly, use your privilege to confront them in that moment. The confrontation does not have to be aggressive. "Hey, Ken, I believe Alicia was speaking. Do you mind if she finishes her thought first? Thanks." Or, you might reach out to Ken afterward, one-on-one, and let him know that while it may not have been his intention, he frequently spoke over Alicia, and that you'd love to see him try to change his behavior.

Additionally, there are tons of resources on bystander training. The global movement Right To Be (formerly Hollaback!) offers resources on how to be an active bystander and intervene in conflicts, depending on the situation and your comfort level (https://samk.app/idc/02-11/). They teach intervention through what they call the five Ds:

- **Distract.** Start a conversation with the person being targeted. Change the subject to take the attention away from the person causing harm. This could also mean physically standing between the two parties in case there is a physical need for intervention.
- **Delegate.** Get help from someone who has the power to make a change, such as a meetup organizer or manager at work, and let them know what you saw and what you need help with.
- **Document.** If it's safe to do so, take screenshots, record behavior, or write down an events timeline of what you

encountered. This can be used for yourself or for others, and can then be shared with the appropriate parties. (If you record video or images, make sure the person who was targeted knows, and ask what they'd like you to do with that documentation.)
- **Delay.** If you're unable to act when something happens, especially if everyone froze during the event, you can always check in with people afterward and ask if they're okay or need support.
- **Direct.** Speak directly to the person causing the harm, or ask the person being harmed if you can help in any way.

You can utilize some of these tools to help support others dealing with discrimination, harassment, or other unpleasant and hostile experiences in the workplace or other design spaces. In later chapters, we'll discuss additional ways of supporting coworkers.

THIS IS JUST THE START

There are going to be plenty of instances where we feel discouraged in this type of work. We may get it wrong along the way, and that's okay *if* we try to do better by addressing our assumptions and biases. We might be discouraged that change is not happening, or not happening fast enough. But remember what we have power over and what we don't, what we can change and what we can't.

Part of why I wanted to write this chapter was to encourage working on the self, because that's a locus of control, unlike larger, systemic issues that can feel overwhelming. When we do the work to improve ourselves, we improve society and culture—there is power in numbers. The changes we make affect how we show up at work, in the classroom, and for our friends and families.

The work of building inclusion starts with the self. By regularly reflecting on our own biases, guilt, growth, and mindsets, we equip ourselves to be more intentional as we start to unpack new ways of being and doing. The next crucial phase is looking at what we might need to *un*learn. In the next chapter, we'll continue the self-work by addressing and dispelling myths that began in our formal education.

3 REVISITING DESIGN EDUCATION

Design, we have a problem. What if I were to tell you that the same "-isms" we explored in the last chapter permeate our industry, and start very early in our relationship to it? We need to examine the root of how we learn about design—in classrooms, online courses, books, and websites.

In my years of being a student and an educator, I've found that institutions of all kinds in North America have some degree of white supremacy and patriarchy at their core. Let's break those terms down. *White supremacy* is the belief that white people are better than others, and that they inherently deserve the institutional and social systems that give them power. *Patriarchy* encompasses beliefs and systems that prioritize male identities and male power over others.

White supremacy isn't always as obvious as a Proud Boys rally, and patriarchy isn't only about male bosses sexually harassing female assistants. These assumptions pervade our society and come out in our behaviors in the form of biases, microaggressions, and gatekeeping. White supremacy and patriarchal systems have, in varying degrees:

- informed what we consider to be "important" or "influential" in design history, including the practices and techniques that we're taught in our field;
- glorified the merit of post-secondary degrees to the detriment of alternatively educated and career-changing students; and
- tainted art and design artifacts, as many of them were acquired through colonization.

Colonization and exclusion are deeply ingrained in design education—but we can change this. In this chapter, we'll discuss opportunities for inclusive practices in design education (with examples from various course curricula) with the goal of reforming how subject matter is taught and empowering students of all kinds through advocacy, opportunities, and skills.

CHANGING THE APPROACH

My first job out of school was at an advertising agency. I struggled to adapt to QuarkXPress because I had been taught to do page layouts in Adobe InDesign. My design education had focused on the specifics of the software rather than how to think critically about my work. Lessons about print processes and the role of electronic tools would have better prepared me for rolling with changes in professional settings.

Students are often taught how to memorize but rarely how to be curious, explore, unlearn ingrained practices, and adapt to change. It's essential that they be taught to think critically in their design practice—pondering the *why*, not just the *how*. bell hooks believed that a person who thinks critically has a more remarkable ability to transform their lives; that thinking critically provides meaning and a sense of agency; and that educators need to do more to teach critical thinking and skills instead of teaching the kind of entitlement that can come with institutional learning: the misguided belief that degrees will give students everything they need (https://samk.app/idc/03-01/, video).

As educators, we can counter institutional barriers to education by changing how we teach. Learning, assessment, research, creation—these are terms we must redefine in order to reach the culturally diverse place that hooks speaks to. We need to reimagine design through the lens of liberation, not just equality. We need to model inclusive ways of thinking about design and teach critical thinking to students so they too can learn in exploratory ways.

The trouble with groupthink

Comic Sans is the butt of many jokes. Originally designed by Vincent Connare for Microsoft in 1994, this font resembles the lettering found in the speech bubbles of comics (https://samk.app/idc/03-02/). However, design students quickly learn in academia: hate this typeface, or you're not a serious designer.

Comic Sans came bundled with most operating systems at the time it was released, making it available to almost everyone. Design that's more specialized or exclusive is often seen as more desirable, while the everyday, accessible, and affordable doesn't feel as unique. But blanket disdain, even of "just" a typeface, leads to the exclusion of certain groups, such as users with low literacy.

Studies show that text set in typefaces like Comic Sans can help with reading comprehension and make reading easier for people with dyslexia (https://samk.app/idc/03-03/). Writer Lauren Hudgins wrote in *The Establishment* about her sister's experience with dyslexia and Comic Sans:

> *The irregular shapes of the letters in Comic Sans allow her to focus on the individual parts of words. While many fonts use repeated shapes to create different letters, such as a "p" rotated to make a "q," Comic Sans uses few repeated shapes, creating distinct letters (although it does have a mirrored "b" and "d"). Comic Sans is one of a few typefaces recommended by influential organizations like the British Dyslexia Association and the Dyslexia Association of Ireland. (*https://samk.app/idc/03-03/*)*

Groupthink, like industry scorn for Comic Sans, begins in design school and persists beyond. Groupthink can negatively affect users who aren't the typical design school graduate. Kat Holmes, author of the *Inclusive Design Toolkit*, wrote:

> Most designers end up using their abilities and experience as a baseline for their designs. This problem is even more pronounced for the predominantly young and able-bodied designers that work in technology. The result is products that work well for people with similar abilities and resources but end up largely excluding everyone else. This is especially true for roughly 1 billion people on the planet with disabilities. (https://samk.app/idc/03-04/)

Besides being immature and likely unfair, unified dislike for a typeface can set the stage for binary design thinking—it's supposed to be one thing and not another; it's good or bad, right or wrong. This binary approach ignores context, history, and non-mainstream ideology, making design seem like a cookie-cutter field where everything is sterile and similar, when, in reality, it should be a field as rich and varied as its practitioners.

Teaching how to learn

One day, while preparing for the Responsive Web Design course I was teaching, I was trying to assess which tool would work best to make images load for the screen or device they were on. Unsure of which resource to recommend to my students, I asked Twitter, and author Ethan Marcotte suggested I try Picturefill (https://samk.app/idc/03-05/).

Picturefill worked perfectly for me, and I eagerly added it to my lesson plan. But my demo fell apart when I tried to show the students what I'd practiced. I panicked.

Rather than continue trying to make the demo work, I showed my students the Picturefill tweet. "This is what I would do at a nine-to-five design job if I didn't know something," I told them. "I follow the steps in tutorials or look at Stack Overflow. If I see tweets or posts shared by other designers about something I don't know, I follow the links or try to find out more. Why don't we click on the Github link and go through it together as a group?"

Suddenly, I had gone from panicking about the demo to showing them how to learn. I was teaching them that they didn't need to know everything; I was teaching them how to be resourceful about finding information, asking questions, and solving problems.

The process of teaching by posing questions is called the *Socratic method*. As a teacher, I found it particularly handy when students worked through problems like coding errors. When they'd ask a question about the error ("Where can I find what I did wrong?"), I'd respond with a question ("Where do you think you can check?"). We'd repeat this dialogue format of answering a question with a question until they figured it out on their own.

Working through the problem, rather than giving them the quick solution, is how we can equip students to handle big challenges in the future. This builds their confidence and teaches them critical thinking. It's a great way to prepare them for careers where self-management skills are key.

Embracing nontraditional paths

The traditional classroom isn't the only place where teaching and learning occur. Sometimes, you don't even need a classroom, just a good internet connection. Many designers are self-taught, learn from trade schools or certificate programs, or earn a traditional degree in a different subject before changing careers. Yet, the design industry—as individuals and as an institution—seems biased against any path that isn't a four-year university or technical-college education that comes with a certificate or diploma.

This academia-only mentality is toxic and not founded on any real data. A fun fact to back this up: my students who took one twelve-week design-and-code course at a coding bootcamp had more contact hours with "design" through lecture, lab, and practice—over 440 hours in total—than my university students who got roughly a total of 120 hours per course, per semester. The topics covered in the bootcamp were equivalent to a few traditional core courses at a university and rounded out to nearly the same amount of time.

The environment for nontraditional paths is engaging. For coding bootcamps, students can take classes in person or online, full-time or part-time. They can set their own pace for their education based on what's going on in their lives. Some can financially afford to take a full-time course and focus 100 percent on their learning. Others may need part-time classes due to jobs, family, or other responsibilities. Coding bootcamps don't take four years to complete and cost four years of tuition, and they're a more viable option for some. The same applies to self-taught learners, and a lot of incredible, nontraditional resources for learning exist online.

This bias toward traditional paths perpetuates the idea that design has one rite of passage. But who decides that the best education comes from traditional academic institutions? That may be true for licensed roles such as doctors or architects, but design doesn't require a license.

Design exists in so many people's lives. Who are we to turn someone away? Imagine how much better our field and practice would be if, like our audiences, we had different perspectives. Plenty of people can graduate from a design program without ample design practice, and plenty of people can be resourceful, design-adjacent, or simply passionate while finding the education, information, and resources they need to succeed on their own. How we get here doesn't matter—what we know does.

As an industry, we need to make education less about *where* people learn and more about *how* and *what* they learn. We can start by reviewing what we teach.

REFORMING THE CURRICULUM

Educators often encourage their students to read their sources with a critical eye. Let's do the same thing when we're designing a curriculum. We can't keep teaching the same material out of habit, because tradition demands it, or because it's too tedious to go through a curriculum-development process. Chances are the curriculum is out of date anyway, either in topic or scope—digital media and web curricula tend to age quickly.

While there may be more enormous hurdles at work, including curriculum review at the college or state level, it's never too soon to start evaluating how we supplement our curricula and what students have access to. We need to look for the gaps in current curricula, build a more inclusive and cross-cultural version of art history, and find ways around institutions so that nonacademic perspectives and resources can be a part of the process.

The curriculum of the past

It has been some time since I, and maybe you, have been around a design program. Let's start by looking at typical four-year design programs, which I went through in the mid-2000s and taught (from 2008 to 2014) before moving directly to nontraditional education.

At some universities or colleges, design programs exist within an art school. Students are taught how to use design software, explore foundational concepts like layout, color, and typography, and undergo periodic portfolio reviews. Like their fellow art students, design students often take drawing, painting, ceramics, photography, and art history classes. Starting around 2010, some programs added courses focusing on user interface and web design.

Art and design history classes are one of the few environments where we learn how art, design, and culture connect. In my experience, however, those history classes focus on a small aspect of art and design history: art celebrating empires, colonization, and European perspectives. The movements taught in most design programs—Modernism, Postmodernism, the Swiss Typographic Style, Gestalt, Bauhaus, and Constructivism—are all based on European culture. Commonly used texts, like *Meggs' History of Graphic Design* by Philip Meggs, focus almost exclusively on white European men, with only cursory mentions of design in other cultures.

Design exists around the world, yet our curricula don't reflect that. Some classes may have a specific European focus, such as the art of Italy and the Medicis or Dalí's influence in Spain. But I've never seen courses covering pan-African art (except in the context of colonization), nor courses specific to Indian, Chinese, or Japanese art. Museum curator and writer Kimberly Drew wrote in her book *This is What I Know About Art*:

> I had learned about buildings all over Italy and done research papers on the Dutch Masters. I was ready to study contemporary art, a field that I thought would be more representative of my interests. Lecture after lecture and week after week, I would diligently research the names that we'd learned—hoping, praying that some of them would be the names of Black people.
> (https://samk.app/idc/03-06/)

Drew's experience is familiar. As a student and someone who has lived in colonized places, these classes made me feel uneasy, strange, and unrepresented. Art and design history focused on the work of the colonizers, while Indigenous people were frequently othered. It was discouraging and alienating; I wondered if art school was the right place for me. It was as though artistic taste and talent were a European export; to have a relationship with art, I felt forced to negotiate a relationship with Europe.

As an educator, I am furious that such narrow perspectives were at the center of my design and art education. I also feel the guilt of having continued those narratives as I tried to fit into my first education jobs. New instructors often teach to curricula they're given, either because they don't realize what they can change or because they're afraid to make waves—and so the old lessons get passed on. What's more, the review of curricula with government agencies is slow and at the whim and internal politics of faculty and departments—if they even want them reviewed.

In the case of coding schools and nontraditional paths, less history is included since more students are focused on technical skills. Training is less academic and a lot more hands-on and practical. Topics may include color theory, typography, layout design, UX theories, and other design fundamentals, before diving into software and code. Students following nontraditional paths don't have to go through supplemental courses like foreign languages, math, or other fine arts classes. It also usually takes less time for regulatory bodies to review a curriculum if it needs to change. In my experience, most US state governments allow us to update curricula to a certain percentage without needing to resubmit for approval because they're aware of the changes required in technical fields.

Flat Iron School has a fifteen-week course that breaks Product Design into five stages covering UX theory, UI theory, usability testing and components, communication and interaction, accessible HTML and CSS, portfolio building, and career development (https://samk.app/idc/03-07/). These topics correlate directly to the workplace. They don't cover the formal design and art history surveys we find in schools. However, even when we teach UX theory or accessible HTML, they're still related to some history, and finding the context is essential. Even letting students know that some of the first computer programmers were women like Ada Lovelace (https://samk.app/idc/03-08/) would shock them given the makeup of the industry today.

Diversify design history

As a student, I could name more of Massimo Vignelli's preferred fonts than I could female designers. A whole world of design is out there that isn't being taught in our institutions. A design education that incorporates a range of influences and identities will enable students to create work with cross-cultural perspectives and design for users who aren't like them—all while preventing students with marginalized identities from feeling othered in their classrooms.

Through access to the internet and a more connected world, educators have so much available to help them expand on design history. Educators already have the tools to build a global, inclusive curriculum by covering topics like:

- Advertising and marketing outside of Europe and America, such as the work of Bobby Kooka and Umesh Rao for Air India (https://samk.app/idc/03-09/), or India's long-running Amul Butter campaign (**Fig 3.1**) (https://samk.app/idc/03-10/).
- The intersections of social and political movements with design, such as W. E. B. Du Bois's use of infographics to tell

Fig 3.1: A 2022 Amul butter ad that features the online game Wordle. (https://samk.app/idc/03-13/).

---

stories about Black American lives in *Data Portraits* (**Fig 3.2**) (https://samk.app/idc/03-11/).
- Feminist design, such as the work of the Guerrilla Girls (**Fig 3.3**) and female designers and artists like Hilma af Klint, Susan Kare, and Paula Scher.
- Queer design history, contemporary designers, and conferences. A great place to start: Queer Design Club (https://samk.app/idc/03-12/), an online directory of queer designers across the world. It was founded by John Voss and Rebecca Brooker, two brilliant communicators who also organize a conference called Queer Design Summit and create space for more queer representation in design.

REVISITING DESIGN EDUCATION 45

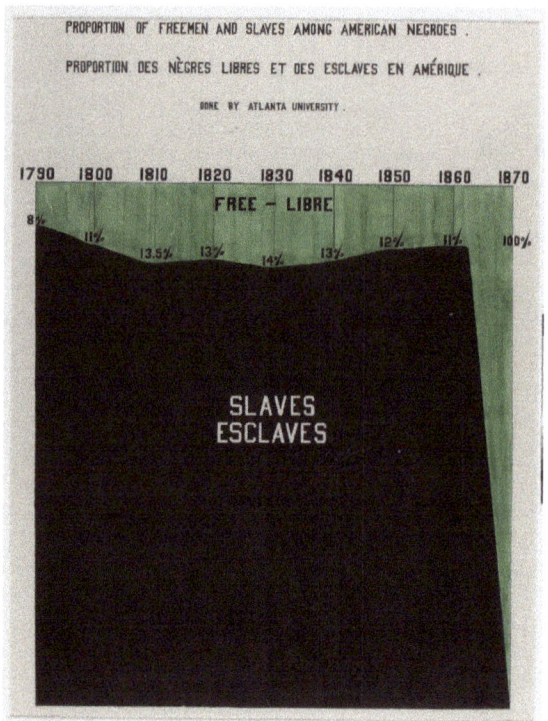

Fig 3.2: Du Bois's infographic chronologically shows the proportion of free and enslaved African Americans from 1790 to 1870 across the United States.

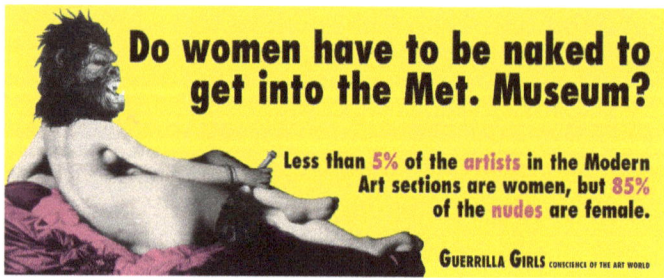

Fig 3.3: The statistics cited on the 1989 Guerrilla Girls billboard haven't improved as of 2005 (3 percent and 83 percent) and 2012 (4 percent and 76 percent). Image copyright © Guerrilla Girls, courtesy of www.guerrillagirls.com (https://samk.app/idc/03-14/).

- Design accessibility, including color contrast, signage, urban planning, and ways that human bodies of all types interact with the world.

These are just a few examples that, if added, build toward a more inclusive curriculum. Design education is pretty homogenous when it's not riding the wave of trends—all the more reason for students to be exposed to marginalized perspectives and histories. Access to these points of view often determines what kinds of designers and thinkers these students will become. It prepares them for the world ahead and builds confidence that those who are marginalized will eventually see themselves in design.

It's also a great way to begin conversations in learning environments—whether online or in a classroom—about the relationship between design and marginalization, as well as design's current focus on appeasing the majority rather than including everybody.

Keep the curriculum flexible

In 2011, the same year Ethan Marcotte published *Responsive Web Design*, my design department asked me and a colleague to review the curriculum for our Interactive Media courses so they could submit updates to leadership at the university and state levels.

The courses—covering basic HTML and CSS, Flash animation, and "other topics"—were outdated. Flash had long been on the decline after the introduction of newer tools and devices (hello, Apple) in the late 2000s and early 2010s. To replace Flash, I suggested we introduce responsive web design, an up-and-coming topic that would give students a competitive edge. The department was open to the idea, and we created a new course focused on responsive web design and other web design topics. The course started as an elective, then became a requirement. It still runs today—with almost the same description we wrote in 2011:

Interactive Media II. *(3-3) Advanced Web Site design, construction and User Interface design. Builds upon the concepts and techniques learned in ARTC 3307. Explores advanced Web authoring with Content Management Systems, design for multiple platforms and advanced Web typography.* (https://samk.app/idc/03-15/)

In this course, we wanted to explore other ways of creating content (through using content management systems), teach responsive web design (though we don't mention it by name), and explore advanced web typography. Nowadays, the course might include topics such as variable fonts or static site generators. The description we wrote has stood the test of time.

It's generalized, theory-centric, and skill-centric, allowing new instructors to adapt assigned readings based on what the industry is looking for at any given time. It allows them to consider new devices and use cases, teaching more inclusively and without being stuck to technologies like Flash or a specific content management system such as WordPress.

For most universities, the bureaucracy of curriculum review can slow down the ability to bring in the latest and greatest information to students. We knew it would take years to update the curriculum again, so we kept course descriptions as vague as possible. By considering the curriculum review process when writing course descriptions, we carved a space for emerging topics and allowed for flexibility in future years.

EMPOWERED CLASSROOMS

To create inclusive educational spaces, we also need to focus on how we interact with our students.

Part of that includes being vulnerable as an educator, as hooks explained:

> *Engaged pedagogy does not seek to simply empower students. Any classroom that employs a holistic model of learning will also be a place where teachers grow, and are empowered by the process. [...] In my classrooms, I do not expect students to take*

any risks that I would not take, to share in any way I would not share. (https://samk.app/idc/03-16/)

What does well-being, vulnerability, and wholeness look like for teachers? Starting points include engaging with "the learning edge" and a growth mindset, both mentioned earlier and easily shareable with students. Additionally, it's beneficial to mention that it's okay not to know everything and to share what we do know. As teachers, discussing our own experiences of marginalization may help build trust with students.

Let students know your intentions for the classroom, that you want it to be a trustworthy and open space where harassment and disrespect have no place. Learning spaces should allow students to bring their full selves into the classroom. You want to do what's best for them, and oftentimes that means advocating for their needs.

Learning environments for all

In both in-person and remote environments, educators can take some easy steps to make the learning experience smoother and more engaging for everyone in the class. One of the best ways an educator can make a space feel safe and welcoming while promoting learning is by considering access to education. Access can be hindered by many things that educators can't see—students may have ADHD or dyslexia, they may not be comfortable with the language used to teach and communicate in the classroom, they may be hard of hearing or deaf, they may have slept poorly the night before, or they may have disabilities but no institutional support for learning with those disabilities.

A classroom experience has its design process in which we want to think through who the users (students) are and what kind of designed solution (classroom) will help them achieve their goals. We can't just design classrooms for the best-case scenario; we need to create the best classroom scenario for all students.

No matter what kinds of experiences, abilities, or access you think your students may have, these steps will help create a more inclusive classroom environment for everyone:

- **Slow down.** Pause for a moment or two between sentences, especially instructions, if needed. Slowing down helps all students catch what is being taught. Whether they're wearing a cast that slows their notetaking, or the course is taught in their second language, make sure to pace lessons for everyone. If you're working with translators or interpreters, there may be a slight delay between what you're saying and the interpreter's translation. Signing takes time to process and comprehend. Be patient with the interpreter and the student(s). An interpreter will usually communicate if they need you to repeat something.
- **Encourage front-row seating.** Try to seat students using interpreters or hearing aids, as well as neurodivergent students, at the front of the classroom. This removes some stimuli and distractions by moving other students out of their direct view. It also means that those using interpreters have a clear line of sight to both the instructor and the interpreter.
- **Share your materials early and often in different ways.** If you're using slides, provide a printed or digital version to students. This supports learners who want to take notes, follow along, or go at their own pace. You might also ask other students to get involved by sharing their notes with the class or in small groups. Research alternative versions of printed books such as audiobooks or ebooks. You might consider sharing curriculum and assignments ahead of time and in multiple formats. Not only does this help students who may need information sooner, but it can also help students wanting to move ahead to continue feeling engaged.
- **Use live captioning or transcriptions.** If you're teaching remotely, enable captions if they're available through the presentation tool you're using, or use a transcription service that can be added to your meeting or educational software. Captions can help with retention while assisting students who can't hear you. Some software plugins include transcription processes; you can also hire transcribers for more accurate text.
- **Address students, not interpreters.** Address students who have interpreters as you would any other student in the

classroom: speak directly to them and maintain eye contact. Avoid asking the interpreters to say something to the student for you, and do not discuss the student with the interpreters. In some states, discussing the student's progress with an interpreter is against the law.
- **Be willing to get creative.** I once taught a class on HTML and CSS in which some of my students were deaf or hard of hearing due to being injured in combat overseas. The students, their interpreters, and I had to work as a team to translate HTML and CSS into plain English and American Sign Language, and vice versa. We made up a language we dubbed HTSL: HyperText Sign Language. We came up with new hand signals for some HTML tags, and they spelled out CSS lines each time, including punctuation. It may seem unconventional, but we were translating among four languages, and through our collective creativity, we made it work.
- **Set a respectful tone.** There will be times that students will disrespect or misinterpret one another. Most institutions have a code of conduct for the classroom. (If yours does not, check the Resources section to start building your own.) The classroom is a sacred space for students, so inclusive instructors should set the tone, reinforce the conduct, and hold students and themselves accountable.

What I didn't expect from many of these accommodations was how much they helped not only the students who requested them, but the rest of the class too. Many students had been hesitant to ask me to slow down, or didn't realize how much better they'd focus if they sat at the front of the class. Removing obstacles for some removes obstacles for all.

Inclusive source materials

Let's look at another way of enhancing inclusivity in the classroom: your source materials. Think of your assigned textbooks. Are they all from textbook publishing companies? Do the authors and editors all share the same identity? Commonly, textbook companies print what their largest share of customers want. Changing textbook publishers' priorities usually requires collective action, including reaching out to the sales representative from one of these publishers and asking for change.

Textbooks can work against inclusivity by being prohibitively expensive for many students. Meggs' History of Graphic Design retails at over seventy-five dollars—add that to the cost of art supplies and design software, and that's a whopping financial barrier. The first book I ever made into a textbook for a course was Ethan Marcotte's *Responsive Web Design*—which wasn't intended to be a textbook for a course. Still, it did predict precisely what would become a common practice over the next decade and beyond. It was available in print and digital formats for under twenty-five dollars, making it an affordable option for students. The pricing made it more appealing for me to include.

Formally published resources also tend to uphold dominant narratives and overindex on privileged identities. By introducing sources that haven't been through academic review processes, we're exposing students to voices that haven't been heard historically in design education contexts. These can also show students how to think critically about the media they consume, what to look for in resources, and how to identify new problem-solving methods.

The web is a living, breathing example of design evolution where nontraditional sources abound. Independent videos, blog posts, zines, podcasts—the list goes on, and these resources are often quickest to embrace new conventions, evolving technologies, and current events. These topics don't always make it into the "historical record" of published texts. Many vital movements around the world, such as Black Lives Matter or the Arab Spring, started on social media. Just because a source isn't formally published doesn't make it invalid; it makes it powerful and genuine. These kinds of sources also tend to be inexpensive or free.

Departmental leadership may be more inclined to choose academic resources related to design because they're considered traditional. However, we know nontraditional sources exist for design, so sharing examples of those sources—and explaining how they place young designers on the pulse of industry advancements—can help make a case for updating the curriculum.

Nontraditional sources for classroom use don't have to go through university review. However, we teachers should evaluate all materials to make sure they are inclusive and thoughtful.

Helping students demand more

In 2019, three Black students at the University of Texas School of Design and Creative Technologies wrote an open letter to the university demanding changes to how the school treated Black students (https://samk.app/idc/03-17/). Their requests included increased financial support, diversity reports and town halls, antiracist learning environments, additional Black faculty and staff hires, and expanded outreach for prospective Black students.

The letter worked. The school responded with a list of pledged changes (https://samk.app/idc/03-18/). It's proof that students have the collective power to change their educational experiences.

At Texas State University, Omari Souza, an assistant professor of design, saw something similar after students asked for more conversations about race in the field of design. Souza took their requests and built out an annual event called the State of Black Design to talk about the lack of Black history and culture in design work (https://samk.app/idc/03-19/). Souza told me that the first event, which was meant to serve about two hundred students, ended up having over four thousand online registrants, more than half of whom attended live. Contemporaries, students, and faculty shared the stage in remote panels to discuss, educate, and learn.

The State of Black Design has become an even more significant event now, and includes a career fair to connect students of color directly to employers. Souza has seen an increase in student job placement, with some receiving internships in their first year of school. If more educators created and facilitated opportunities for students the way Souza did, more design students would be better off.

Unfortunately, students are taught early that institutions have all the answers—even when those institutions do little to provide them. It takes just one person like Souza to change the game. Encourage students to take control of their education whenever possible. When students feel they have agency, they become better students.

A CALL FOR CHANGE

Inclusivity can reshape design education—and in turn, the design industry—but only if we notice the limitations for some and make learning accessible for all. It's time to demand *more*—of our institutions, our ways of learning, and the spaces in which learning happens.

We've established that design education exists in many places and that there isn't one right way to learn. We've discovered how to rethink our approach within design education and its topics. We've discussed better ways of structuring content in group settings like classrooms. Our instruction prepares students for the design industry—in a big way. If we start making these changes in the education phase, their impact ripple out into the industry.

All of these elements—history, representation in design, and the way we learn—don't end with education because we continue to learn on the job. An investment in design education is an investment in the design industry.

4 TRANSFORMING HIRING PRACTICES

There is nothing I get asked about more than hiring—and I mean *nothing*. It's an area that companies with design departments need the most support and help with.

Here's how many hiring processes unfold in our industry:

1. Companies slap a diversity-related paragraph on their job posting, publish the posting on a few channels, and then wait for applications to come in.
2. They assume that "diverse" candidates will apply for these roles because their company reputation makes them appealing.
3. When candidates with marginalized identities apply, leaders and hiring managers pass them over for various reasons:
    a) They assume that the candidate isn't as capable of filling the role as someone with a dominant identity and that hiring them would be "lowering the bar."
    b) They worry that the candidate wouldn't "fit in" with the company culture once they're hired.
4. When the company (inevitably) doesn't reach its hiring goals, they say there aren't enough diverse people in the industry to apply—the "pipeline problem."

This process, while common, has zero potential to result in better hiring practices and diversified teams. Often, when marginalized folks experience some of these situations, they consider whether they need to withdraw, reject an offer, or tell others. If we want to change *whom* we hire, we'll benefit most by advancing *how* we hire.

In this chapter, we'll focus on creating a thoughtful hiring process that welcomes and encourages candidates with diverse experiences. This process will improve internal dynamics as well, highlighting the benefits of planning and communicating goals to the rest of the team. The most challenging pivots are those we make to our teams and company culture, so we need to take time to research the problem well—as any designer would.

HIRING MYTHS

Okay, real talk: How many times have you heard that hiring processes are complicated for everyone involved? What if I were to tell you that although hiring is complex, some marginalized identities deal with *even more* hurdles to get to—and go through—the hiring process? When I think of systems of oppression, I think of prejudiced banking practices and redlining, education systems banned from teaching about race and history, exploitative carceral and legal systems, and even biases in professional recruitment.

Recruitment may not have the same widespread community impact as the others, but it is still a system with all sorts of oppressive practices baked into it—and it needs a big reset.

Our first step is to unpack the many myths about hiring practices in tech. Let's return to the example hiring process we went through and break down all the myths companies might buy into.

"Lowering the bar"

Many people in leadership believe that increasing diversity means "lowering the bar" or reducing the quality of hires. This argument only holds if you believe that candidates with marginalized identities are less competent than their dominant-identity peers. This profoundly prejudiced belief is ubiquitous in our industry and in others.

Yet there are countless examples of managers who hold this belief, either overtly or subconsciously. In 2015, Twitter's senior vice president of engineering Alex Roetter said during a meeting: "Diversity is important, but we won't lower the bar" (https://samk.app/idc/04-01/). In 2021, Snowflake's CEO Frank Slootman said: "We're actually highly sympathetic to diversity but we just don't want that to override merit. If I start doing that, I start compromising the company's mission literally" (https://samk.app/idc/04-02/).

These leaders (and many others) give in to stereotypes that all marginalized hires will perform poorly. They may be making assumptions about these hires' quality of work or only supporting candidates from specific schools, companies, or institutions they know about. Other ways we see leaders excluding marginalized identities is through nepotism, or hiring family and friends rather than unknown candidates. It says a lot about the leadership and how they view their team dynamics—specifically people with marginalized identities, whom they consider "less than."

Hanna Naima McCloskey, CEO at Fearless Futures, wrote about this harmful myth:

> *There is an explicit assumption that there is an irreconcilable clash between talent and the proposal of hiring minoritised and marginalised people. [...] This phrase sets up whatever the default practice for hiring and promoting people is as by definition fair and just.* (https://samk.app/idc/04-03/)

McCloskey's words illustrate that stereotypes, prejudice, and "-isms" are based on perception, not facts. She went on to point out that our society's normative standards are "white, middle class, heterosexual, able-bodied, cis, and male." By repeatedly leaning on this norm, we're perpetuating a perspective—a prejudiced one—over and over, keeping it in place although it holds no truth.

Perhaps we need to state the obvious. For the CEOs and leaders who believe that increasing diversity means lowering the bar, what they mean to say is that they want dominant identities to stay in power, and that they're not interested in broadening their company culture. They're also admitting that they don't see merit in people who don't look, act, or think like them—which calls to mind the myth of "culture fit."

"Culture fit"

In the 2010s—when in-office kegs or happy hours were a common perk companies boasted about—tech employers would often say that their metric for whether a candidate was the right fit for their team was if they wanted to get drinks with that person. If a candidate could kick back and have a beer with the team, casually known as the "beer test," it was a good sign—sometimes carrying more weight than a candidate's actual ability to do the job.

But a "culture fit" based on drinking is inherently exclusive (not to mention potentially harmful). What if a candidate is pregnant? What if they suffer(ed) from an alcohol addiction? Or have a gluten sensitivity, or take medication that precludes drinking, or just don't enjoy alcohol? Not only would they be unable to participate in workplace activities once hired, but they'd also be put in the uncomfortable position of having to disclose personal health histories or religious practices. As a critical note: never ask people why they aren't drinking—it's rude and, in some cases, illegal.

This mindset-turned-practice sends a very particular message about what a company's culture looks like—and who fits into it. "You end up with this big, homogenous culture where everybody looks alike, everybody thinks alike, and everybody likes drinking beer at three o'clock in the afternoon with the bros," said human-resources consultant Patty McCord in the *Wall Street Journal* (https://samk.app/idc/04-04/). While not all "culture fit" is about drinking beer, this example illustrates that tests for culture don't account for people's unique experiences and backgrounds, instead aiming to make sure people check a box or fit in with everyone else.

I believe companies want to hire someone good to work with. So, how do we define "good to work with"? Kind? Funny? Quiet? Busy? Inspiring? Maybe. What most companies and teams are looking for is someone who enhances the team by bringing something new to the table. That could be previous work experience, parts of their identity, where they've lived, or apps they've worked on. Diverse teams are more innovative, productive, and do better work (https://samk.app/idc/04-05/).

"The pipeline problem"

The "pipeline problem" is a common, overused excuse to explain the lack of diversity in an organization's applicant pool. This myth places the blame on applicants rather than on companies and their lack of internal policies that would attract applicants instead of excluding them. Examples include hiring from the same well-known schools (which also have exclusion issues) or the mindset that the company's reputation is good enough to attract diverse employees.

There *is* no pipeline problem. There are plenty of qualified candidates out there for any position. However, when those candidates don't apply for a role, most companies assume they don't exist rather than asking *why* candidates aren't applying.

The reasons range from gender or racial pay gaps, to being ignored for promotion, to toxic and unsafe work environments, to biases about who seems right for the job. There's also evidence of bias in hiring that excludes people due to their name, photo, race, education, and more. Marginalized designers and their allies use private channels to discuss unsafe and exclusive hiring processes or experiences within a company; word gets around.

If we keep recruiting students or designers from the same schools and competitors, we'll run into many of the same issues. If we keep insisting that it's the candidate pool and not us, we risk perpetuating our bias toward our own company. The process can always be improved, even if it means simply changing where we historically and repeatedly recruit from.

Exclusion in hiring

The above myths come down to fear of the unknown, change, or inconvenience. Yes, the hiring process is a vulnerable, time-consuming, and challenging experience for all parties involved. And yes, hiring is frequently rushed, with little opportunity to step back, slow down, and reconsider the approach. Fixing it isn't easy work, but it is crucial work.

We need to start thinking through how to make hiring work for everyone—candidates and hiring managers alike. A better hiring process may not only reduce harm to candidates with marginalized identities, but may also attract more candidates, build better teams, improve internal participation, and change company culture. Revising the process now can save time, money, and headaches in the long run.

Now, let's tap into our growth mindset and peel back the layers of the hiring process.

PLANNING FOR THE ROLE

When roles—especially existing ones—open up, it's usually because of an immediate need: filling a vacant seat, meeting headcount goals, or supporting workload. But if we can sidestep the urgency and look at the hiring need with genuine curiosity, we'll be able to better—and more inclusively—frame the role.

Whether there's an immediate hiring need or not, hiring managers need to strategize pre-emptively, the same way we do in other parts of our roles. We need to form clear goals and assess what's missing on our existing teams before starting the hiring process for any candidate. That way, we can work toward creating a strategy based on the current needs.

Define your goals

Before posting a role or starting to recruit, identify the goals you have for this new hire. Spend time thinking through your ideal candidate and what they could bring to the table:

- Where might they be coming from? Is it straight from school? A code camp? A similar role at another company?
- Are they a career-changer who can bring in a wealth of contextual knowledge from another industry?
- How did they learn the skills to make them qualified for this job? What assumptions can you make about their educational experience?
- What skills could they bring that your team lacks?
- What experiences outside of design could they bring to the role?

These are great discussions to have with team members, other hiring managers, and human resources (HR) colleagues. Getting perspectives from as many team members as possible can help you identify missed opportunities, learn about challenges in the process, and gain insight into why people have left the role in the past.

Assess your team

We tend to unconsciously hire candidates whose skills and backgrounds resemble those of our existing team members—or, for that matter, ourselves. This is called *affinity bias*, and it's another reason why "culture fit" can be a dangerous metric for recruitment. Focusing on what the team is missing instead of what's familiar is one way of avoiding homogenous hiring.

As a team, discuss the following questions:

- What skills are firm on your current team? What are you lacking?
- What skills are you missing that a candidate can bring in?
- Are existing team members vocal about new roles? Do they actively recruit for the team?
- What does an inclusive team look like? What do you want the team to achieve together?
- What accommodations do you provide for interviews (e.g., sign language interpreters, reserved or reimbursed parking, hotel or flight reimbursement)? How can you share this with a candidate?

- Does your team have the necessary training to review portfolios, facilitate interviews, and provide candidates with appropriate accommodations? Have they been trained to notice and reduce bias in themselves?
- How can you support candidates who may necessitate more complex hiring processes (such as immigrants or enrolled students)? If you're not open to those candidates, what steps can you take to include them?
- What have you learned from opening roles in the past? What would you do differently?

While there is often pressure to fill roles quickly, taking this initial research step will make implementing the process go more smoothly. It will inform teams and align them on hiring goals. It will also encourage the habit of asking questions about all the possible experiences a candidate could bring in, removing the expectation of that one ideal candidate—who, frankly, doesn't exist.

SCALING A PATHWAY

At one point in my career, I directly managed about thirty instructors at a code school across twenty-one locations, and had over thirty roles open for additional instructors. We were growing exponentially and needed to hire fast and intentionally.

In those frantic days, many hires worked out, and some didn't. Most hires made it through a term or two; others returned to the world of individual contributors. We realized that our hiring process—inspired by an iterative "figure it out as you go" startup culture—wasn't scalable for this large talent pool. None of our interviewers were on the same page; they often didn't know what the end-to-end hiring process looked like (especially since it was changing), nor who was responsible for each step. It also wasn't clear if everyone was assessing candidates similarly. The assessment was subjective, and although candidates often had experience with mentorship because most were coming from the industry, they didn't have direct classroom experience.

To ensure that interviewers were informed, confident, and consistent, we needed a game plan. So I set out to document the objectives, goals, and tasks for each stage of the hiring process to empower the interviewing team to make decisions.

Map your hiring process

Interviewing requires coordination, communication, and clear expectations—not just between the candidate and interviewers, but within the organization itself. A mentee of mine once went through a hiring process twice with the same company, completing the same (unpaid) interview tasks with the same people because they didn't have a plan. Candidates can tell when the hiring process is disorganized; they may remove themselves from consideration, and spread the word to their network.

Mapping out the hiring process was my first step toward getting everyone aligned, organized, and aware. Taking inspiration from the Google Ventures Design Sprint exercises for mapping (https://samk.app/idc/04-06/) and thoughtbot's Critical Path exercise (https://samk.app/idc/04-07/), I created a visual representation of the whole hiring process (**Fig 4.1**).

Next, I created a detailed interview process table that listed the goals for each stage, the people involved, the candidate assessment, and the hiring status (**Fig 4.2**). To identify the people involved, I used a Responsibility Assignment Matrix (RAM or RACI) (https://samk.app/idc/04-08/) to identify who was responsible for each stage, who was accountable for completing which task, who needed to be consulted for decision-making about a candidate, and who needed to be informed before moving forward.

This table not only ensures that team members are aligned on objectives, candidate assessment, and ideal outcomes; it can also increase awareness of bias (intentional or unintentional) in the process.

When hiring designers and developers, the focus is often on portfolio or code reviews—but the best interview processes get to the core of who the candidate is and what they might bring to the organization. Mapping out the entire process from

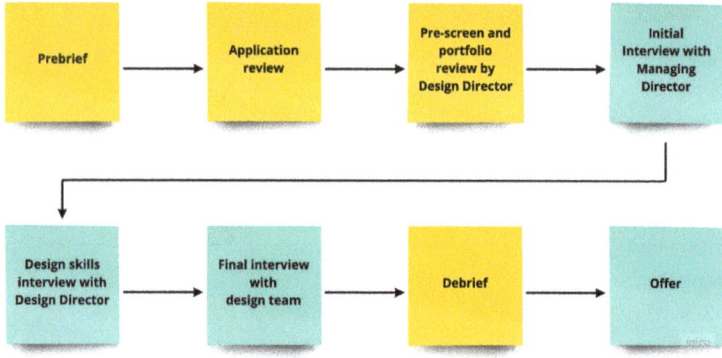

Fig 4.1: An example of a critical path for the interview process. This exercise is a thoughtful way of framing the entire interview journey. The yellow steps are internal; the blue steps include the candidate.

pre-screening to onboarding ensures that you and your team see the process holistically.

Schedule safeguards

Pre-briefs are internal meetings that happen before interviews. Hired.com recommends them for all interview processes:

> *The pre-brief meeting is exactly what it sounds like: a briefing before interviews begin in order to ensure everyone is clear on their role within the interview process. If candidates wind up answering the same questions in multiple interviews, not only will they get fatigued, but your team won't get a well-rounded, accurate assessment of their abilities.* (https://samk.app/idc/04-09/)

Pre-briefs are an opportunity to share goals and assessments from the first few timeline stages during a hire. Sharing perspectives in open discussion is an effective way to engage team members in the process, learn from them, and identify ways to mitigate bias.

Internal Status	Pre-screen and Portfolio Review	Interview #1: Intro and Process	Interview #2: Design Skills	Pairing/Final Stage	Debrief	Offer
Goals	Identify core requirements, opportunities. Confirm visual design and code through prior work	Learn more about the candidate including how they align to company values and what they bring to the table	Pair on critique and coding challenge	Pair on design challenge	Discuss and review finalists	Comp approval from managers and HR
Responsible	Design Director	Managing Director	Design Director	Design Director	Design Director	Managing Director
Accountable	Design Director	Managing Director	Design Director	Design Director, Design team	Design Director,	Design Director, Managing Director
Consulted	N/A	Design Director	Managing Director	Design team	Design team	Design Director, Managing Director
Informed	N/A	Design Director	Managing Director	N/A	Managing Director, HR	N/A
Assessment	Scorecard	Scorecard	Scorecard	Scorecard	N/A	N/A
Hiring Software Status	Applied/Initial Review	Non-technical	Technical	Final Stage	N/A	Offer

Fig 4.2: An example of an interview process table for hiring a designer.

Like pre-briefs, *debriefs* enable more consistent and fair evaluations of candidates. They are retrospectives where all team members involved can share what went well, what didn't, and define what the next steps will look like.

During debriefs, diversity and inclusion consultancy Paradigm recommends that you:

> ... *add structure to the discussion (pro-tip: center the discussion on the specific attributes you were assessing for, and ask each interviewer to give their perspective on that attribute), ensure every interviewer shares their perspectives, have more senior people speak last to avoid anchoring effects, and call out potential bias.* (https://samk.app/idc/04-10/)

Debriefs are a nonnegotiable stage for all interviews, not only out of fairness to the candidates, but also as a safeguard for ensuring alignment and improving processes. Some companies ask candidates to complete questionnaires about the interview experience; this data can provide excellent insights for debriefs and future pre-briefs.

While pre-briefs and debriefs require more meetings, they're worth it to ensure your team is on the same page and able to pivot when necessary. Hold these meetings for each candidate if you can, but if that's not scalable, at least conduct them at the beginning and end of every open role.

BETTER JOB POSTINGS

I once saw a startup community account complain on social media about not having enough applications for their startup accelerator, which aimed to provide support for female founders (the pipeline problem myth in action!) (https://samk.app/idc/04-11/). They shared a link to the accelerator, hoping more candidates would apply, but when I opened the link to investigate, I immediately saw problems:

- The application used words like "secret meetings" and "exclusive," which gave off a "boys' club" vibe—and likely

turned marginalized founders away (https://samk.app/idc/04-12/).
- The company offered support through one-on-one meetings with the board, which (surprise, surprise) was made up entirely of white men. As a woman, this surprised me and made me worry about being in a room alone with someone I didn't know.

Later, through research and speaking with others, I learned that some of these board members had been accused of harassment or retaliation by their employees—proof that one-on-ones would be unsafe for marginalized founders.

When I responded with suggestions to address these concerns, the accelerator owner became angry and defensive. Our mutual followers encouraged him to meet with me to hear more of what I had to say, but instead of attending our meeting, he sent another team member. The team member insisted their approach was correct, told me I didn't know much about the startup space, and demanded proof about the board members in question. In other words, it didn't go well. No changes were made, and they continued to recruit fruitlessly for a while. Luckily, there are other accelerators that focus on looking for and supporting a diverse group of founders, such as DivInc (https://samk.app/idc/04-13/).

If we're looking only at our hiring needs, we're ignoring the needs of the people we're recruiting. If our language or processes alienate candidates, the negative impact far outweighs our intentions.

A job posting is often a candidate's first meaningful interaction with an organization, and what it includes (or doesn't) can determine whether they even decide to apply. Granted, writing an inclusive job posting can be a challenging task. How do we ensure it communicates the role accurately while also appealing to potential applicants?

Inclusive language

When creating job postings, write clearly and as neutrally as possible to appeal to various candidates. In job postings from

Webflow, *voice and tone* help make their roles more appealing and inclusive. They use a more casual, direct, first-person tone, paired with encouraging messages such as:

> If you don't meet 100 percent of the above qualifications, you should still seriously consider applying. This template takes into account that candidates from marginalized groups may think themselves that they aren't qualified for the job—because of existing bias in the industry. Studies show that you can still be considered for a role if you meet just 50 percent of the role's requirements.

Remember, you're setting up your candidates for success, not trying to overwhelm them with challenges or waiting for them to mess up. Some more recommendations for improving your job descriptions:

- Avoid words like "ninja," "rock star," and "unicorn"—all way too common in design and development spaces. Not only are these words gender-coded, but they also don't clearly define anything about the role.
- A tone that skews too businesslike can come across as cold and off-putting. Corporate jargon (like "highly qualified") probably won't mean much to applicants. Try to be more specific and conversational.
- Address people directly in the posting. Use "you" instead of "the candidate" or "the applicant" so it sounds like you want to hire an actual human for the job.
- Ask colleagues—especially those from other teams or departments—to review your job descriptions and point out exclusive language. The more people who review the job-posting language, the better.

In my early days as a design director at thoughtbot, I found that many of our job postings could be improved with more inclusive language. I used an AI tool called Textio to audit our existing role descriptions (https://samk.app/idc/04-14/). It gave

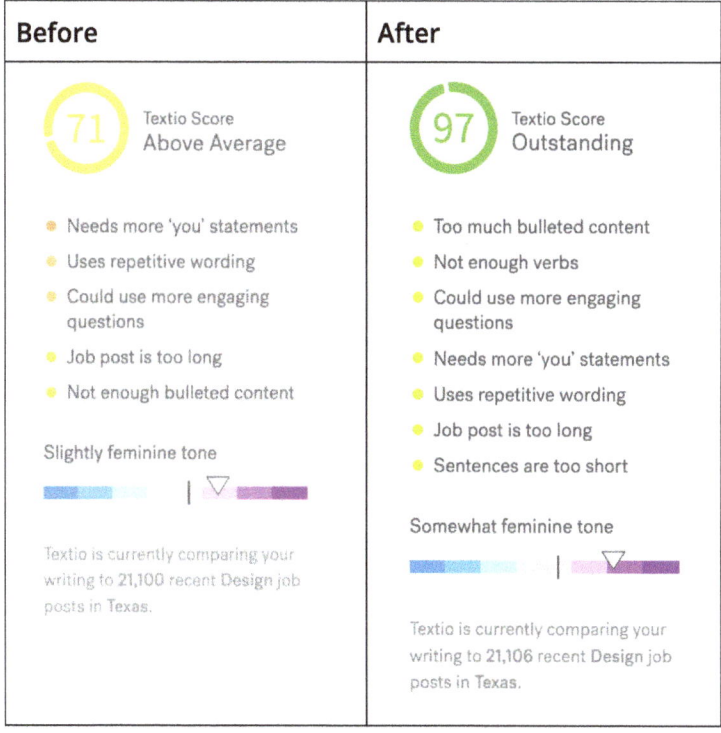

Fig 4.3: On the left is the Textio score for our original job posting. On the right is the Textio score after we made their suggested updates.

---

me great suggestions for improving tone and removing gendered language (**Fig 4.3**).

After implementing the suggestions for our designer and design director postings, I ran the descriptions through Textio again. The first score went from 71 to 97 (out of 100), and the second from 85 to 94. Most importantly, once I opened the role, I saw greater variety in skill, age, talent, expertise, and gender among the candidates who made it past the first round—and more applications than I had seen before. More women applied, as well as more designers older than the industry average (early thirties.)

Transparency

Transparency is an essential ingredient for writing more inclusive job postings. The more information we provide about the company and its values, the expectations for the role, and the interview process itself, the more candidates and interviewers will be aligned beforehand, leading to more effective recruitment and interviews.

Transparency can also be a brilliant way to share more about company culture. When you're clear about your company values, you attract candidates who find those values appealing. For example, Bumble's job postings help set expectations for candidates while also making their values clear:

> *We strongly encourage people of color, lesbian, gay, bisexual, transgender, queer and non-binary people, veterans, and individuals with disabilities to apply. Bumble is an equal opportunity employer and welcomes everyone to our team. If you need reasonable accommodation at any point in the application or interview process, please let us know.*
>
> *In your application, please feel free to note which pronouns you use (For example: she/her/hers, he/him/his, they/them/theirs, etc).* (https://samk.app/idc/04-15/)

Project Include, a nonprofit advocacy group aiming to increase inclusivity in the tech industry, also recommends sharing information about what it's like to work at the company (https://samk.app/idc/04-16/). Does the role allow for flexible work hours? Are there expectations for the length of the workday or workweek? Some roles even share what they hope a candidate will achieve in their first thirty, sixty, and ninety days on the job. Giving details about the work pace can help attract talent and improve retention.

Another great way to improve a listing is by being transparent about the salary range. Listing a salary range attracts candidates who already have a number in mind. It also saves time on both ends and avoids unpleasant surprises in an already grueling process. If the company or role is new enough not to have an established salary range, or is open to hearing compensation requirements, say so.

Job postings should also let candidates know what to expect in the interview process, including the number of steps, the hiring timeline, and accommodations available. Recently, I've seen companies outline each stage of the interview process with a sentence describing the task, time, and people involved—for example: "In this stage, you'll pair on HTML and CSS layouts along with the design director, Sam, for about an hour."

These different types of transparencies improve your job postings by setting clear expectations for candidates. Potential candidates with values, needs, or timelines that don't align with yours will be less likely to apply, saving everyone the headache of discovering misalignment later in the process.

Recruiting

Simply hoping that suitable candidates will stumble across our job posting isn't enough. Recruiting means reaching out to talent—not waiting for them to come to us. If our goals are to build an inclusive team, we need to meet candidates where they are. This will take more budget, time, and effort, but it will lead to better interviews and a stronger team.

Step outside of the default places you recruit. Here are a few suggestions for improving your outreach:

- Post to Slack, Discord, and other online professional communities, including social groups for industry conferences. Browse job boards, directories, and resources that cater to marginalized identities. (See the Resources section of this book for specifics.)
- Seek out candidates from community lists and databases that are specifically meant to elevate professionals with marginalized identities, such as Tatiana Mac's Devs of Colour (https://samk.app/idc/04-17/) or Amélie Lamont and Timothy Goodman's People of Craft (https://samk.app/idc/04-18/), which supports designers and developers of color. There are

other directories that cater to specific identities, including gender, disability, and immigration status (see Resources for more).
- Reach out directly to candidates in your network. Write personal messages for each potential applicant and tell them why their specific portfolio and experience make them great candidates for the role. Be careful not to tokenize them.
- Search for candidates on networking sites like LinkedIn. Share your job posting intentionally with an individual and ask if they'd like to speak about it.
- Ask your existing team to informally recruit from their networks, or through conferences or other events they may be attending.

Whenever you reach out to people on any network, be intentional about your requests, and mindful of others' time and energy. Avoid sending copy-and-paste emails and spamming inboxes.

In addition, you can broaden your candidate search to include students and recent graduates:

- Send your job postings to university programs and code schools. They're often eager to share open roles with graduating students and alumni.
- Partner with schools to help students develop the skills related to your open roles. Shopify does this through their Open Learning program (https://samk.app/idc/04-19/); IBM Design Bootcamps put students on three- to six-month hiring paths (https://samk.app/idc/04-20/).
- Reach out to international student centers at universities (most have them) and let them know if your company employs immigrants. You can also post roles within the international student office or email the roles (there is almost always a job board).
- Go to portfolio reviews if you're invited, or reach out to schools in your area and ask their design programs if they have portfolio reviews or senior exhibitions.

When paired with inclusive language and transparency, sharing your job postings in the right places can dramatically increase your chances that suitable candidates will see the role, share it with others, and choose to apply. That said, posting roles in places that cater more to marginalized communities isn't the end of the story. If the interviewing process doesn't mirror the postings' promises of inclusion, candidates *will* shy away. It's crucial to make sure your words and actions align.

BIAS IN SCREENING

Women are 47 percent more likely to be seriously injured in a car crash than men, and 17 percent more likely to die (https://samk.app/idc/04-21/). Why? Crash test dummies were built to represent average male bodies. Differences in male and female stature translate to differences in how close or far our feet are from the pedals, where the back of our car seat has support against whiplash, where seat belts cross over us, and on and on. It's not that male engineers were purposefully excluding female bodies; their implicit bias simply created a fatal oversight.

Bias can lead to painful, traumatic, and exclusionary results in any interview process, especially within screening. Screening is the process of determining if a candidate is suitable to move on to the formal interview stages. Many excellent candidates are dismissed during screening for elements of their identities that have nothing to do with their qualifications for the role. Let's look at some of the most common biases we must overcome when screening job applicants.

Career background bias

A brilliant developer I know began her career as a dog walker. When she was first working on her developer résumé, however, she left out the dog walking because she didn't think it was relevant. I persuaded her otherwise—it had taught her all sorts of relevant skills, such as communicating with clients, self-starting, time management, and budget management.

Candidates with career backgrounds that aren't in design and development often bring new skills and perspectives, including adaptability and a willingness to learn. Discounting them can be a missed opportunity to hire highly motivated candidates with curious mindsets. If you see applicants with unexpected or unrelated roles, ask more about their experiences and how they view their career path. You'll not only be pushing back against bias but also identifying potential candidates with surprising talents.

Collegiate bias

A similar bias exists toward candidates who attended a four-year college. Screening for these candidates means missing out on highly qualified people who are self-taught or who learned their core job skills through bootcamps or online programs (https://samk.app/idc/04-22/).

With so many learning resources available on the internet, self-taught designers and developers aren't uncommon—in fact, many of the designers and developers who follow me on Twitter are self-taught (https://samk.app/idc/04-23/). At the same time, hiring managers often say they shy away from candidates without a traditional college education.

Candidates with a four-year degree may have the primary skills a job requires. But that doesn't guarantee they're the best fit. There are also downsides to traditional education—there's more of a focus on "teaching to the test" or memorizing, which may not always translate well in the workplace in terms of self-management or motivation.

Reframe your thinking about educational backgrounds. Whether an individual applicant is suitable for a role doesn't come down to how they received their training—it comes down to how they use it.

Experiential bias

Coming from an educational and management background, I've heard plenty of myths about the perils of hiring junior candidates. They're overconfident. They can't possibly have the skill set required. Their work will be substandard, and they'll require far more resources—and handholding—than someone who's been out of school for a few years. That's a ton of assumptions based on little more than a graduation year.

In reality, juniors have fresh perspectives and varied skillsets, especially if they're coming in as career-changers. There are quite a few benefits to being open to a junior hire:

- Juniors have fewer ingrained habits to unlearn and may be faster to pick up the processes and values of the companies they join.
- Waiting for your senior-level unicorn can be expensive; it's often quicker and less costly to hire a junior-level designer or developer and mentor them into a senior. They're also more likely to stick around.
- Having juniors on the team gives middle and senior management a chance to grow their communication, coaching, sponsorship, and mentoring skills.
- Mentoring juniors helps non-juniors reinforce processes and techniques in their minds, leading to better communication with clients and stakeholders; the non-juniors become better employees themselves.

Hiring juniors is an investment that grows over time. Ruling out a junior can, in some cases, come across as ageist or showing prejudice. Juniors have a lot of energy, new ideas, and good questions. We're gatekeeping the industry if we don't give them a chance—especially when we have roles we can't seem to fill. The juniors don't need to adapt; we do.

Seek out mentees to work with, or find ways your organization can invest in programs to support juniors. In Chapter 5, I'll share more ways to get involved in bridging the gap between juniors and the workplace.

Citizenship bias

Immigrants are often seen as one monolithic group, but they have vibrantly different experiences. You may see job candidates who are international students, partners of folks already in the US under a work visa, or refugees settling in a new place after doing the same job in their home country. Immigrants have a lot to add, from new perspectives and skills to cross-cultural awareness and empathy. Going through any immigration process tends to impart a keen awareness of systems thinking.

Hiring noncitizens can sound confusing and costly, which tends to scare some screeners away—especially those at smaller companies. It shouldn't. While immigrants and employers may get some extra paperwork, it's completely doable. Immigration sponsorship costs can also be a fraction of what most companies spend on recruitment ($3,000 to $5,000, not counting the length of time the role has been left open) or leaving a role unfilled. That said, applying can sometimes cost from $1,000 to $4,000 for either employers or immigrant candidates (https://samk.app/idc/04-24/).

In many roles, immigrants are automatically screened out of the process if companies don't have processes in place to sponsor them. You may not have control over whether their application can move forward, but you can schedule conversations with HR and managers to learn why. With luck, HR will be open to an initial call with an immigration lawyer who can answer questions about the process. If none of this is possible, there are still ways to support immigrants, even if you can't hire them. Perhaps you can suggest a remote mentorship program, set aside office hours to answer any questions they might have, or refer them to other companies in your network that might hire them.

Working to combat bias

Some companies have found it helpful to temporarily remove personal information like a candidate's name, photo, and educational background from early screening steps, allowing screeners to focus on the application, résumé, and portfolio. I mean, seriously—why are photos even necessary? It's a practice that can introduce more bias, and it's got to go. For example, in *Design for Cognitive Bias*, David Dylan Thomas reported that the city of Philadelphia had experimented with hiding personally identifying information in their applications:

> First, when hiring a web developer, the city learned that the best way to anonymize a résumé, even in the high-tech world of web development, was to physically print out the résumé and have an intern (or someone with no stake in the hiring process) physically redact the personal information with a marker, just like a CIA document.
>
> Second, when the hiring managers found a résumé they liked, they would typically go to GitHub to view that developer's profile, which, when loaded, would immediately reveal all of that applicant's personal info, thus ruining the experiment. So, clever people that they were, they wrote a Chrome plugin that would redact the information as it loaded. (https://samk.app/idc/04-25/)

Many companies have turned to AI and other digital tools to filter personal information out of applications. While AI hasn't worked well to date in screening candidates (https://samk.app/idc/04-26/), browser plugins and other tools can reduce some potential for bias by keeping parts of the candidate's identity anonymous.

But anonymized hiring only gets us so far. After all, the interview process can't stay anonymous forever—and that's where bias can start to creep in. According to the Forbes Human Resources Council, employers have a "natural tendency […] to hire for culture fit rather than culture add" (https://samk.app/idc/04-27/). When we set off with biased expectations for a role, we're likely to end up hiring for those expectations, rather than being open to different approaches and candidates.

Ultimately, we can't rely on software to make candidate decisions, nor can we rely on our instincts. Hiring on autopilot is both ineffective and dangerous; we must revamp how we think about screening and actively work to remove personal and systemic biases from the equation. Ditching these old habits means ditching the things that make our interview processes inequitable and exclusive.

INCLUSIVE INTERVIEWS

I once conducted a remote interview meant to be a design-pairing exercise with screen-sharing. The candidate had an unreliable internet connection and joined the virtual meeting by phone call. When we realized the misunderstanding, the candidate and I felt flustered, and I didn't have a backup plan. This made for an awkward and rocky interview.

In workspaces with pricey computers and high-speed internet, we forget that candidates may not always have access to what we do. They may live in rural communities without enough internet bandwidth for video calls. They may not have access to transportation to and from interview locations. They may not have high-end digital equipment at the beginning of their design or development careers. Time zones, work or home situations, or disabilities may limit how and when candidates can communicate.

Interviews are the process by which a candidate and a company can get to know each other better. An inclusive interview process ensures both parties are set up for success and have everything they need to make that process happen.

Accommodations for all

One of the areas companies are least prepared for is supporting applicants with disabilities. Disabilities may be physical or mental, visible or invisible, short-term or lifelong, incidental or congenital. Many candidates choose not to share their disabilities with interviewers to avoid discrimination. Others may need accommodations for the interview process.

According to the US Equal Employment Opportunity Commission:

> *Employers are required to provide "reasonable accommodation"—appropriate changes and adjustments—to enable you to be considered for a job opening. Reasonable accommodation may also be required to enable you to perform a job, gain access to the workplace, and enjoy the "benefits and privileges" of employment available to employees without disabilities.*

> **An employer cannot refuse to consider you because you require a reasonable accommodation to compete for or perform a job.** (https://samk.app/idc/04-28/)

Whether a candidate has disclosed a disability or not, a company must be able to accommodate disabilities at any stage of the process. Companies should have these accommodations in place *before* they're needed; otherwise, they risk elongating the interview process for the candidate and themselves. It's also important to make it clear to candidates how they can request accommodations for their interview. My employer, Netlify, says it right on their Careers page:

> *What if I need an accommodation for the interview? No problem! If you need accommodations for the interviews, please let your recruiter know or contact accommodations@netlify.com.* (https://samk.app/idc/04-29/)

Have a plan for all interview needs and do your research ahead of time. There are a few links in the Resources section to help you get started, along with EEOC guidance on what you can and can't legally ask a job candidate.

Facilitating communication

I am not—by any means—an expert on disability. Much of what I know comes from the ADA (Americans with Disabilities Act) website. It offers excellent advice that can apply to anyone who strives for inclusive communication in interviewing.

Here are a few things I've learned about facilitating this type of inclusivity:

- **Communicate early and often.** It doesn't hurt to have everything spelled out and explained early on. Go over the interview steps from start to finish (as discussed earlier with job postings) so candidates can make plans or requests accordingly. Share lists with the names and roles of everyone candidates will talk to during the process. Consider providing a

written (printed or digital) copy of the interview questions for the candidate in advance so they'll know what to expect.
- **Create opportunities for candidates to request accommodations.** Many online hiring tools have questions related to accommodations listed in their application templates. Ensure that any other teammates involved in the interviews are aware of any necessary accommodations.
- **Be flexible about location.** If you conduct on-site interviews or require full-day or multiday interviews, ensure that candidates know the time commitment as soon as possible. Share transportation and hotel options, offer to make travel arrangements, and reimburse any out-of-pocket costs (or let candidates know what financial support they can expect). Purchase or reimburse candidates for a day pass at a coworking space or help them reserve a meeting room at their local library. Some candidates prefer to interview from their own home, but others may appreciate a more private space with better access to equipment.
- **Arrange for different types of interaction.** Seat interviewers next to each other so candidates who rely on lip-reading can read efficiently. If working with sign language interpreters, seat the interpreters next to the candidate they're interpreting for; you want seating arrangements that help the conversation flow. Also, consider captioning or transcribing the interview to share with the candidate afterward.

You can find more information about accommodations in the Resources section of this book.

HIRING IS THE EASY PART

While the content in this chapter may suggest otherwise, hiring is the easy part. Getting into an inclusive mindset, reflecting on what the team really needs, figuring out how to interview in a way that sets everyone up for success—this all happens *before* anyone is committed.

It's just the start. Think about it: it's a lot easier to get someone through an interview process than to make sure they feel included at a company. In the next chapter, we'll look at those steps: making offers, supporting employee retention, and creating stronger workplaces.

5 LEADING INCLUSIVITY IN THE WORKPLACE

As of 2021, only nineteen out of the 1,800 Fortune 500 CEOs are Black (https://samk.app/idc/05-01/). That means 1 percent of all Fortune 500 CEOs are Black, while the most recent US Census data shows that almost 15 percent of the population identifies as Black (https://samk.app/idc/05-02/). That's a considerable gap.

As we've seen, bias in classrooms and hiring processes can create obstacles for people with marginalized identities. Once they're brought into a company, the culture can either continue this trend or promote inclusivity to better support and retain its team members.

Retention means different things to different people. At a high level, retention refers to the steps a company takes to keep its employees. That can be measured by how long they stay, how happy or fulfilled they are, and more. Retention is often seen as time-consuming and expensive, but hiring is more so. It's in a company's best interests to hold onto the talent it hires.

The responsibility of setting the cultural tone is on company leadership, but employees at every level have the power to improve the workplace for themselves and others. Even if you can't change certain processes, you can advocate for others, ask questions, and model behavior. In this chapter, we'll look at some factors that impact a company's inclusivity—either convincing employees with marginalized identities to stay or sending them on their way.

SALARIES

The tech industry has a pay disparity tied to gender, race, and ability. Recent data by the World Economic Forum in its Global Gender Gap Report, published in March 2021, is sobering (https://samk.app/idc/05-03, PDF). While laws supporting equal pay exist in some countries, the report shows it will take an average of 135 years for most countries to close the pay gap among genders.

The design industry isn't immune to this inequality. After reviewing data from the 2019 Design Census (https://samk.app/idc/05-04/), InVision shared that female designers made 73.1 percent of what their male counterparts made (https://samk.app/idc/05-05/).

This is clearly an issue in our industry today. Pay is inequitable and unfair, and there's a stigma around discussing design salaries. To reach pay equity, we need to challenge these patterns together.

Equitable compensation

The best way to begin closing the pay gap is to be transparent about salaries and ensure that everyone is paid the same rate for the same work. Being clear about salary ranges—and building out rubrics to assess placement within those ranges—can reduce bias in the process. As Project Include puts it:

> *Create clear salary/equity bands with defined expectations for each to create a level playing field and reduce bias. Be transparent about them to fight the gendered and racialized pay gap, which can increase over career lifetimes. [...] While there may be no direct link between pay secrecy and pay inequality, pay secrecy appears to contribute to the gender gap in earnings.* (https://samk.app/idc/05-06/)

If your company is looking to create an inclusive workplace, equitable compensation is the place to start. Here are some recommendations you can take to senior leadership or HR:

- **Define what factors play into determining pay bands.** This will help you create a fair and documented formula for determining salaries. Research how location, inflation, cost of living, skill level, and tenure are calculated for role- or team-specific pay bands.
- **Audit existing salaries.** Once you create a new formula for pay bands, assess existing salaries against it. Compare demographic data between people in the same or similar roles to identify where pay gaps exist. Document the findings and bring them to the appropriate managers for implementation. Build out clear communications about each stage and what the team can expect as a timeline.
- **Spotlight the issue.** Let the team know why you're auditing, then work on individual communications about raising salaries where they need to be raised.
- **Train managers on how they're expected to review and apply the new formula.** Consider having HR audit new formulas moving forward so more checks are in place to avoid additional disparities.
- **Maintain transparency about salaries.** Buffer makes its employees' salaries public, along with its approach to compensation and its salary calculator (https://samk.app/idc/05-07/). If you can't be that public at your own company, see if you can publish salaries internally, or at least within your team. For open roles, share salary bands so applicants know what to expect.

Salaries are often hard for anyone outside of HR or senior leadership to control. However, not all hope is lost. There's excellent information online about how other companies, such as eBay, have worked on pay disparity. And there are still other ways to get the conversation started.

Talking about money

Some managers may cringe at the thought, but I encourage employees to talk openly about their salaries. There's a taboo around discussing pay—especially in the workplace—but there shouldn't be. Transparency about salaries and benefits is one of the few tools employees have to promote equitable compensation in their companies.

If you aren't able to audit or change salaries, you can still help shift the needle toward equity in your workplace by using one or more of these approaches:

- Reach out to your peers with similar roles and similar levels of experience to see if they'll chat about salaries, both in and outside the company.
- Ask your managers for more salary transparency. Ask if salary bands for all roles exist and if they can be published. Reinforce the idea that transparency around salary keeps people happier, keeps them in their roles longer, and attracts new team members. It's important to note, too, that countries like the US legally protect employees who ask about or discuss salary; employers cannot stop it.
- Share what other companies are doing around salary equity and transparency—especially in the same field—with a trusted manager, on open company forums or internal channels, or by directly asking HR. If you're up to it, write a proposal or present your findings in a slide deck. (I find that these formats can communicate things more clearly.) You can even reference examples like Buffer. Lastly, connect with like-minded employees within your organization or team that can support you, especially if they're senior level.
- If internships at your company are unpaid, pull together information about why that's not a fair labor practice and urge senior leadership to start paying workers fairly at every level of experience.

Pay is not something everyone can control. That said, open conversations with peers and leadership within your company may help move things in the right direction.

NONTRADITIONAL BENEFITS

Gone are the days of employees seduced by kegs in the kitchen or haircuts at their desks. While such perks might seem appealing in the earlier stages of a career, they're trendy and fleeting—and don't contribute to professional or personal fulfillment. They aren't benefits that will retain employees over time.

People want benefits that will enable them to do good work and pursue bigger life goals. That may go beyond retirement plans and health insurance—a needed standard worldwide. Nontraditional benefits are a key part of an employer's offer to a candidate, enticing them to work for—and stay with—the company that supports their life.

Caregiving benefits

According to a report by Pew Research Center, the US has some of the least friendly parental-leave policies for employees compared to forty other countries. Countries like Estonia, Japan, Norway, Germany, and Korea have a government-mandated minimum number of weeks for parental leave. Estonia is currently at eighty-six weeks of paid leave—more than a year and a half! Japan and Korea set aside more paid leave for new fathers in 2019: thirty weeks and fifteen weeks, respectively (https://samk.app/idc/05-08/).

The US isn't so lucky. The federal government doesn't mandate leave policies; until it does, it's on employers to provide caregiving leave for their employees. And let's just say they could be doing a *lot* more.

Comparatively generous employers in the US might provide a few weeks of parental or medical leave, but never enough to cover the time needed for significant medical treatments or recovery from childbirth; nor does typical leave cover inclusive definitions of parenting, family, caregiving, or medical needs.

Companies have an opportunity to accommodate and value employees by offering leave policies that support:

- **Various birth-related situations.** Natural births, adoptions, surrogacy, miscarriage, and sudden guardianship merit the same level of importance and paid leave benefits. Shockingly, to this day, it's rare to find employers who give miscarriage-related leave.
- **Non-child caregiving needs.** Some employers provide benefits for taking care of family members, pets, or other dependents, like roommates, neighbors, or grandparents. Caregiving benefits could include pet insurance or additional paid sick leave. In many marginalized communities, there are unique and extensive networks of care that may include people outside of immediate family, so coverage for them is essential.

These types of benefits are more inclusive of the different types of lives people lead. They don't make assumptions about what a family or birth should look like. Employees are more likely to stay with a company that shows this kind of empathy.

Flexible work

When COVID-19 brought most of the world to a halt in 2020, tech workers were largely unaffected (if they weren't laid off), continuing their projects remotely. Some companies have adopted flexible work-from-home policies since then, while others have made a permanent shift to remote work.

Remote-first companies can attract working parents with caregiving responsibilities, those seeking a healthier work-life balance, and those with far commutes, high transportation costs, or long shift hours. Work flexibility can help keep many women in the industry who've had to quit during the pandemic to take care of their families and other dependents. At the same time, don't assume that one way of working is a catch-all. Many people like going into the office, needing in-person collaboration or a quiet setting that their workspace at home just can't provide.

Make how, when, and where people work an attractive benefit by letting them choose. If they want to work from home, don't use spyware or demand cameras be turned on all day;

trust people to be professional, and supply them with the necessary office equipment (standing desks, ergonomic chairs, dual monitors, etc.). Facilitate the environment that lets them do their best work—on *their* terms.

Additional benefits

While plenty of benefits exist, some can be deliberately used to counter systemic challenges. Benefits that offer financial or community support—as opposed to benefits that feel more like luxury items or one-time perks—can help build a more inclusive workplace. Here are some examples that support employees' well-being:

- **Professional development.** Everyone benefits from access to continued education, including courses, workshops, training, conferences, books, time to invest in personal projects, and more. With the rate and speed at which design trends and tools change over time, having access to continued education—supported culturally and financially by employers—is essential. Supporting employees' professional development tells them you are interested in their growth and want to support them in their careers.
- **Loan matching.** One newer benefit that's gaining traction is loan matching, where employers match contributions to pay off employees' student loans. Loan matching is inclusive because it acknowledges other parts of people's lives that might otherwise keep them worried or stressed, helping to alleviate the problem.
- **Transportation or support for remote days.** Some employers provide parking, pay for public transit, supply indoor bike storage, or allow days to work from home so a commute isn't necessary. I've always particularly liked seeing companies offer parking and public-transit support. First, this potentially gets cars off the road, reducing pollution—climate change impacts marginalized communities the most (https://samk.app/idc/05-09/). Second, it supports alternative ways for people to get to work if they need to commute.

- **Work immigration paperwork fees reimbursement.** In some cases, immigration fees are covered by the company in others, by the immigrant. Either way, the company benefits greatly, yet it's a costly out-of-pocket expense for the immigrant. Consider providing reimbursement for some of the costs related to the long process of immigration applications and visa processing: legal fees, application fees, photo fees, the gas cost for transport to the nearest immigration office or embassy, passport fees, doctor and medical test fees, etc. No other employee has these out-of-pocket costs; the burden is on these immigrant workers just because they come from outside the country.

This isn't an exhaustive list, but it can help us start to look at how we can counter systemic issues that our teams and communities might face. Find opportunities within the company to implement as many of these as possible—without lowering anyone's salary.

LEADING THE WAY (FROM ANY LEVEL)

While only high-level managers can control salaries and benefits, anyone can lead the way in making their spaces more inclusive. Communicating issues, advocating for change, and supporting others is leadership, no matter your role or title. Let's look at how we can create more inclusive spaces for our coworkers and colleagues, starting with meetings.

Better meetings

Meetings are not inclusive or comfortable for everyone invited. The design and tech industries both have a pretty bad reputation for holding meetings where attendees with marginalized identities are silenced—typically by white men taking advantage of systems that only uphold their voice. We need to be aware of these dynamics and consider how we can make meetings work for all attendees.

Inclusive meetings happen when all attendees can value the thoughts and opinions being shared. There's an expected level of respect and accommodation for all participants, and it's honored throughout. Whether you're leading a meeting or attending one, the responsibility of inclusivity is the same, especially if you're *not* from a historically marginalized group.

Here are a few recommendations for making meetings more inclusive:

- **Respect schedules.** Start and end meetings on time, even if other attendees are late. When scheduling meetings, consider remote attendees' time zones and pick business hours that work for everyone, sending reminders as needed. Provide agenda items and meeting goals beforehand, so attendees know what to expect and can prepare if needed.
- **Provide accommodations.** We discussed accommodations for interviews, and those same recommendations hold here. Recognize that different people retain information in different ways. Notes, captions, transcriptions, and recordings of meetings can help neurodivergent team members and those with disabilities.
- **Don't require cameras on.** On video calls, encourage team members and peers to turn off their cameras if needed. People have a variety of reasons why they can't or don't want to be on camera, including needing a bite to eat, having a choppy internet connection, not wanting to distract other attendees with something in their environment, and other reasons. Normalize cameras being off; treat video calls like phone calls. Trust your attendees—they're adults. If cameras are needed for lip reading, invest in sign language interpreters.
- **Speak up when witnessing interruptions.** If others are being interrupted, interject and say something like: "Excuse me, Sandra was speaking, and I'd love to hear her finish her thought. Sandra, please continue." Do this in real time rather than waiting until later, if you can.
- **Give credit where it's due.** If someone co-opts someone else's idea, name the person who originally stated the idea. Repeat if necessary. While we're all on the same team and

move ideas forward together, it's essential that marginalized or quiet team members get credit for their work—it can help support their career advancement and push past biases.
- **Rotate administrative housekeeping** Assign rotating note-takers and facilitators for repeat meetings, ensuring marginalized team members, especially women of color, aren't always given this chore. In *Harvard Business Review*, Joan C. Williams and Marina Multhaup wrote that the problem with "office housework" is that it tends to fall on people with marginalized identities without contributing to their career trajectory:

 > Office housework happens outside of the spotlight. Some is administrative work that keeps things moving forward, like taking notes or finding a time everyone can meet. Some is emotional labor ("He's upset—fix it."). Some is work that's necessary but undervalued, like initiating new processes or keeping track of contracts. [...] [These tasks aren't] tied to revenue goals, so they are far less likely to result in a promotion. (https://samk.app/idc/05-10/)

Some or all of these tips may apply whether you're in a remote workplace or at the office. The bottom line is that you want to help everyone feel heard, respected, focused, and included in meetings. When everyone is given the space and respect to participate in the discussion, it allows individuals to show up as their best selves.

Leadership audits

Whenever companies ask me for advice about building diversity on their team, I ask what their leadership representation is like. I loosely call the conversation that follows a *leadership audit*. Its goal is to learn about the makeup of identities in senior leadership and the roles they hold, company-wide goals, and current approaches to inclusion in the hiring, retention, promotion, and support of employees. Through leadership audits, we can get to the core of the issues we're trying to solve and quantify them.

Not to mention that learning about leadership representation often says a lot about the makeup of the rest of the team.

Ask questions

Questions are a good starting point. We want to learn what a sample subset of people in power are thinking, what they're motivated by, and how they work with their management and non-management teams. To conduct a leadership audit, we need to ask ourselves (and our team) these questions:

- What does our leadership team look like?
- Do we have a similar percentage breakdown in gender or race to the general population?
- Who is represented in technical management roles and non-technical management roles?
- Does our leadership represent various demographics or identities? If not, why is that?
- What does our leadership say about our team to potential applicants?
- Why is this important to us?
- What have we done so far? What has worked, and what hasn't?
- What questions can we ask our non-managers regarding our culture in a fair, anonymous, and welcoming way?

When we get more comfortable asking these questions at work and in meetings with teams and managers, it can inspire others to do the same. After all, these are questions middle managers and senior leadership need to be asking themselves. Anyone can bring them up—leaders aren't only those with leadership titles.

Do your research

As you're asking and answering questions, look for other insights to bring into the discussion:

- **Gather demographic data.** Your HR department may have some data available. I recently pulled together data available

to me to see if I could draw any correlations to retention. I used it to hypothesize why people were leaving and what opportunities we had for improvement.
- **Discuss improvements with colleagues.** Schedule one-on-ones with other managers to brainstorm. Learning new ways of working—or iterating ideas based on one another's experiences—could lead to collaboration opportunities.
- **Ask others in your network.** What are their companies doing to bring more representation into leadership? I'm constantly checking in with conference Slack groups, where I know peers in similar roles—or roles that would report to someone like me—are having discussions. We constantly talk, compare notes, and ask each other questions about how we can improve as managers to better equip our teams.
- **Seek out diversity, equity, and inclusion (DEI) consultants.** I've put in requests for consultants in digital suggestion boxes, asked for training from my managers, and called in support (and budget) from directors. The Resources section includes the names of some consultants I've worked with.

Given the state of the tech industry, you'll likely find that your leadership team is not diverse. The next step is to have conversations about changing that.

Start conversations

The most fruitful conversations take findings from research and audits. I've found it effective to recruit others and teach them how to gather this information about their own roles; there's power in numbers. Here are two steps I've asked them—and am now encouraging *you*—to take, independently or collaboratively:

- Request a meeting to discuss your audit's results and opportunities for improvement. Know that the conversation may cause discomfort since company-wide issues will be discussed and (hopefully) addressed.
- Look at and share public data that other companies put out to help convince your managers and leaders that change is possible.

This can be difficult and brave work—difficult because we fear for our jobs (especially when we're from marginalized backgrounds) and brave because we speak up even when we're afraid our concerns will be ignored. That said, bringing things up can encourage others to follow our example.

Many in leadership roles assume that this kind of work will cost them their current position, but I'm not suggesting we remove specific people from leadership. I'm simply highlighting the importance of understanding how leadership teams come to be. What leads people to these roles? What are the requirements they meet, and what potential oversight might they have in these roles? The more we know, the more space we make for future leaders.

In one-on-one settings, you may find it helpful to start these conversations on a smaller scale with people you trust, such as peers and managers. There are anonymous methods for sharing your findings if that's more comfortable. Otherwise, you can always discuss feedback options with HR, employee resource groups, or DEI councils.

DEI initiatives

Companies and employees can also take more direct and intentional action to focus on DEI. Employee resource groups and DEI councils are two common initiatives that enhance belonging and inclusion at a company.

Employee resource groups

Employee resource groups (ERGs) are groups of employees with shared experiences or identities, such as Asians and Pacific Islanders, women and femmes, veterans, parents, and more. ERGs host educational, cultural, or fundraising events, advocate for group members' interests, and create a community for employees. ERGs are created with an intention to provide a safe space to discuss issues unique to the group, and to collectively bring those issues up with the company. Success in carrying out that intention is largely dependent on support from and trust in leadership.

ERGs tend to be more common in enterprise or large organizations, which can support groups with facilities, space, and funding. Most ERGs are housed under the HR department, where they can directly siphon company resources to activities and initiatives.

Ask if you can start an ERG at work—or ask if a proposal and budget can be built out for different groups that would like to start one, even if you don't personally identify with said groups (that's called being an ally). If ERGs already exist at your company, support any events or celebrations they have.

One year, on International Women's Day, I was invited to speak at the Women's ERG at an ecommerce giant's office. While there, I met a male ally who'd become a member to ensure more male employees supported the work of the Women's ERG. He even told me he'd asked each male employee he knew to bring another male employee to celebrate. The rest of the group felt pleased and supported when they saw allies from different backgrounds showing up to the event.

Diversity councils

While they have similar goals to ERGs, diversity councils are often broader—woven into the whole company instead of focused on individual employees or identities. They're one way that many organizations look at improving diversity across all teams: by tasking people already invested in the company to make the workplace more inclusive.

That said, diversity councils vary widely in their structure and impact, depending on the companies they're a part of. I've seen councils draft onboarding diversity curricula, fund nonprofits that train novice designers, launch scholarships, and take ownership of in-company training. Councils can help improve hiring and retention, jumpstart upskilling opportunities for the team, or even build out internship programs. Since inclusion touches almost everything in the workplace, there are plenty of opportunities.

I've also seen diversity councils lose steam and focus after the initial stage. There's usually one culprit: inaction by senior leaders. Council members and senior leaders must work together in writing and reviewing policies, addressing improvements, and making changes that support all types of workers in the workplace. No diversity initiative can succeed if senior leaders don't step up.

There are other challenges, too. Work on diversity councils is compensated at some companies, but not all, and council work can add a lot to people's plates. Employees with marginalized identities are often expected to be actively involved without any recognition or additional pay—and without compassion for the emotional toll diversity work can exact. Additionally, because diversity councils are often organized under HR, some employees hesitate to speak up or get involved for fear of retaliation. Housing such initiatives under operations or the CEO may not only alleviate those concerns, but also better enable diversity work to be tied to all aspects of the company.

Diversity initiatives must also look beyond gender and race. Not all types of diversity are visible or physical. It's essential that company leaders work with diversity councils (and external consultants) to understand how to invest in employees with less apparent marginalized identities (such as those related to mental and physical health, educational background, and socioeconomic status). Consultants can help bridge these knowledge gaps and structure benefits to the employees and company at large.

No cure-alls

I've seen DEI initiatives that have succeeded and many that have failed miserably. The ones that do well have leaders—specifically senior leaders—investing in the work themselves rather than expecting employees to add the work to their jobs or unpack their traumas for the company's benefit.

Leaders need to set the example if diversity councils are to be effective. Expecting frontline team members—who don't manage anyone, don't have decision-making power or influence, or are entry-level—to do this work alone only erodes trust. *People with leadership roles need to model the behavior they want to see.*

Diversity councils and ERGs will not solve all present issues; unfortunately, there's a misconception that they will. While these initiatives can be helpful, expecting them to single-handedly solve an organization's diversity and inclusion issues would be a grave mistake.

Additionally, problems aren't solved overnight; they may take a few months or years to solve. The work isn't easy—and the presence of an ERG or diversity council doesn't mean that a workspace is free of inclusion issues. It just means that there might be a space to work on solutions.

These solutions are meant to assist or help move initiatives along, provide a space for marginalized people to identify with, and instill a sense of belonging in a united front. But everyone, especially managers, needs to do the work consistently and often. Focusing on the bottom line and ignoring people is the quickest and most painful way to destroy inclusivity at your workplace. Success requires buy-in from leadership, changes in mindset, doing the research, owning the burden, and building the practice.

INCLUSIVE MANAGEMENT

When you think of leadership, you probably think of planning or strategizing high-level ideas to help move the company forward. But for inclusive practices, it takes a lot more. Employees need to see that their leaders are doing just as much as they are to make the workplace inclusive. Simply telling teammates to move diversity or inclusion efforts forward is not enough; coworkers will need to see their managers and C-level executives take action. As a manager myself, I believe managers—especially senior leadership—are 99 percent responsible for making their company inclusive.

What makes an inclusive manager? In Chapter 2, we discussed methods for taking action: watching our language, listening actively, and intervening. We've also seen how design education can exclude certain students. We've learned about the hiring myths that infiltrate company culture and the makeup of teams. Actively working against exclusion and myths is part of being an inclusive manager.

But that's not all it takes. Managers hold power, and their actions significantly impact their teams. There's a misconception that if frontline workers take diversity-related action, it will trickle up to the whole organization. But as the ACT Report by the Catalyze Tech Working Group states:

> Tech CEOs and leaders must model inclusive leadership for their senior team and managers at every level. How leaders prioritize diversity, speak about talent, and reward or sanction behavior establishes—and can reset—company norms. Changes at the top flow down and influence the behavior of other leaders, middle managers, and hiring managers who, in turn, translate company norms into hiring practices, interview behavior, and decisions. (https://samk.app/idc/05-11/, *PDF*)

The most important thing I want to tell managers looking to improve DEI is that it starts with senior leadership and then proliferates down. No matter how much work the rest of the company does, if senior leaders don't match the effort, nothing will get done. Inclusivity has to be prioritized as much as any other part of the company.

Being better managers

There's a lot managers can do to become better advocates for inclusion. I mean, don't we all *want* to become better managers? Mark Kaplan and Mason Donovan, authors of *The Inclusion Dividend* (https://samk.app/idc/05-12/), write that leaders need to:

- have a heightened self-awareness in order to audit themselves,
- be able to understand individual and group identities—and build relationships with them (think back to the learning zone from Chapter 2), and

- understand they are likely "insiders" who fail to see how status benefits them while missing how it can cause issues for others.

Awareness of how the "insider" manager identity pairs with other identities is key. The authors write that "leaders are more likely to be engaged when they understand why a change effort is important to them personally and when they can communicate this to others."

I agree with this advice, especially their points on self-awareness and how role dynamics change perspective (which means we need to work harder on getting feedback). Managers are most effective when they build trust, know when and how to apologize, and do not depend on reports to tell them when something is wrong and how to fix it. Here are a few actions that put those recommendations into practice:

- **Standardize performance reviews.** Surveys, rubrics, and evaluations are all viable options. Run through them with directions beforehand or fill them out yourself. This will provide clarity on how the review works. Have the employee self-evaluate, then follow up with your assessment. Talk about the assessment together so you can align on expectations.
- **Don't promote in a silo.** Talk to direct reports about their interest in management or an individual contributor path. Ask them questions about what they hope to achieve and what will help them get there, rather than projecting your managerial expectations on them. While an employee might be a perfect fit for a role, it's best to support them in the path they're most interested in, especially if they're historically marginalized. Provide leadership training, spend time answering questions, and encourage them regularly.
- **Stop asking, "Where do you see yourself in X years?"** I've always found that question very hard to answer, especially during times in my life when I had a temporary work status. My perspective aside, many employees find planning for the future daunting, worried they'll make a wrong decision.

Instead, go with questions that seek more concrete answers to unlock possibilities for your team, such as:
- What are you excited about learning?
- What is something you'd love to teach others?
- What do you enjoy the most about your role currently?
- What's missing in your role?
- What does growth look like to you?
- How do you like to receive feedback on your growth?
- What support are you looking for in your current or future roles?

- **Invest time in your team.** A manager I know blocks time weekly to do extra check-ins with his team, requesting feedback and asking people how they're doing. Sometimes, a little communication about how people are goes a long way in supporting your team.
- **Choose proactive, not performative, action.** It's common to want ourselves or the company we work for to look better. However, many focus on doing and looking good with the least effort (performative allyship) instead of using their power for impactful change. I've seen examples such as senior leaders posting blanket statements about International Women's Day while also having cultures in which women get spoken over in meetings or are paid less than men doing the same role. Performative actions show that managers don't know what they're doing, grasping at straws while doing the very least, whereas proactive actions demonstrate thoughtfulness and managerial foresight.

These are all steps a manager can take to work toward a healthier work culture. However, we don't always have the luxury to be proactive when so much of our time is spent reacting. Most managers will tell you they're prepared to handle issues as they arise, but there's a difference between saying something and doing it.

When issues arise

When designers notice something that can be improved within a project, they're usually the ones to take ownership of that task, especially if it's an area where they can make an impact or get experience. But issues related to inclusion (e.g., microaggressions, harassment, and retaliation) are not the same because they aren't specific to a designer's direct tasks.

Employees who bring up inclusivity issues in the workplace are usually the ones experiencing them—vulnerable people with marginalized identities. And when they bring them up, it's because they're trying to escalate the issue to the people who may be able to solve it. They're looking for leadership to take responsibility. Yet, frequently, those tasks get assigned right back to those employees—and they shouldn't. Employees don't necessarily have the power or expertise to solve these problems, nor should they be taking time away from their projects. Not only that, but if they're personally affected by the issue they raised, the work of addressing it head-on may mean navigating their trauma.

By raising the issue, they already feel the burden of marginalization and have calculated the risks to their career and personal safety. If they've identified an area of valid concern, it's the responsibility of the company (specifically HR and managers) to fix it.

However, many employees are wary of HR. It's a part of the company and therefore not immune to bias and politics. Still, it's essential to include HR if there are safeguards and processes; documentation of the concerns may be handy in the future. This doesn't guarantee action from HR or leadership, but it can create a paper trail.

If you are the one bringing up issues, especially as someone from a marginalized community, find people who will support you, perhaps dominant identities who can vouch for your experience and use their privilege for the greater good. Or find colleagues in similar roles or identities who can support, corroborate, or help you communicate harm done to you.

It's also important to keep records of the situation. Screenshots of chats, emails, and other communications can provide necessary evidence when issues arise. If escalations aren't working (sadly, quite common), enlist others for support. You may have to push a lot, and there's no correct answer for knowing when to save your energy or when to push harder. That will change for each scenario, but it's something to be aware of; this work is tiring and draining.

Finally, remember that anyone with power or privilege in their identities can help address problems. It's crucial (and in some cases legally required) to help resolve or continue to escalate the situation to the right people.

But what about when leadership does nothing? If leadership doesn't start investigating HR complaints or addressing issues, company culture and productivity suffer, and leaders probably don't have a plan to deal with it. Their apathy isolates team members and creates tight workspaces. Employees feel like they're not being heard or may continue being harmed. They may feel cornered, forcing them to leave the toxic environment. No one wins when they leave.

When employees leave

There are plenty of reasons for someone to leave. They may not feel supported by their manager, they may sense they're not growing, they may be looking for a change in role or career, or they may have found something better. What isn't talked about enough is when an exclusive, unsupportive environment forces out employees. Luckily, there are reports we can learn from, like the Ford Foundation and Kapor Center for Social Impact's "Tech Leavers Study" (https://samk.app/idc/05-14/, PDF), which found that 62 percent of employees would stay in their role if the company invested in creating a more "positive and respectful work environment." In comparison, 57 percent said they'd have stayed if the culture was more inclusive. How did we get here if we've tried to hire inclusively?

Consider a common experience for many new hires with marginalized identities, as illustrated by Quebec's Centre for Community Organizations (**Fig 5.1**).

When a new hire is first brought on board, they're warmly accepted in the name of diversity—but things can change quickly. Once they bring up inclusion and equity issues within the organization, they encounter pushback and excuses from management. Management then retaliates, and the employee ultimately leaves the organization. This tale is way too familiar

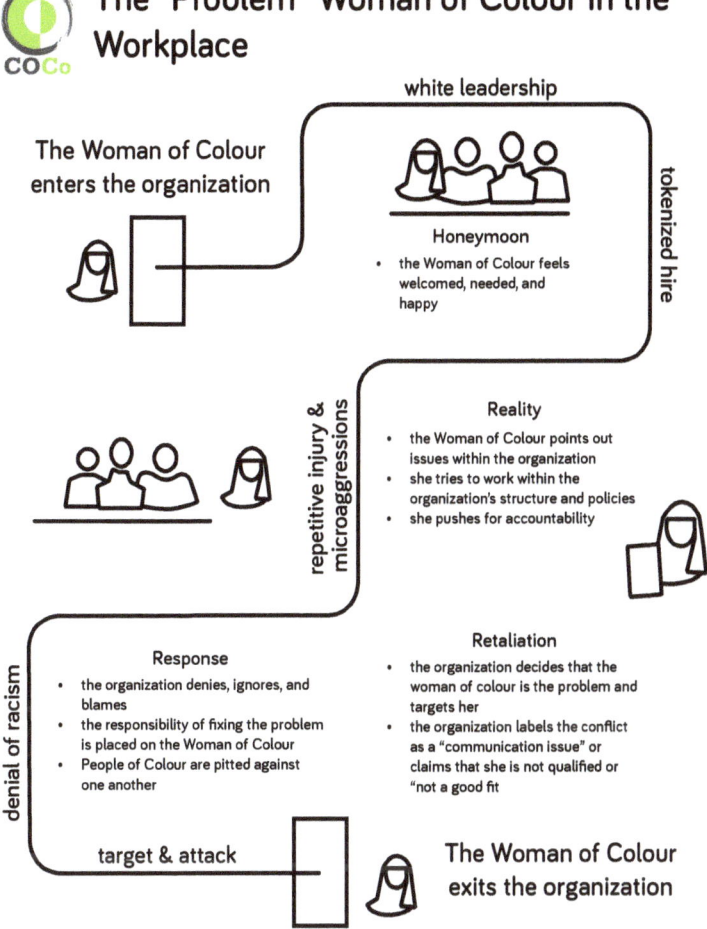

Fig 5.1: Employees with marginalized identities often leave after experiencing their company's unwillingness to hear criticism and make changes. Illustration from the Centre for Community Organizations, as adapted from the Safehouse Progressive Alliance for Nonviolence (https://samk.app/idc/05-14/).

for employees with marginalized identities, and it's why many workplaces have retention problems.

Company culture and safety can only improve if feedback is acknowledged and acted upon. If not, the culture will suffer, and word will get around (internally and externally) about why people are leaving. Retention is a problem that can be mitigated by many practices we've looked at in this chapter: equal pay, good benefits, supportive colleagues, and open-minded leadership.

When employees do leave, companies have an opportunity to collect feedback, identify ways they may not be fully supporting their employees, and prevent the same inclusion problems from harming future employees.

Exit interviews

A good exit interview allows the departing employees to share their real experiences. Why? It's difficult for employees to be open with feedback since many fear it may cost them opportunities, promotions, or even their job. Just because current employees don't share criticism doesn't mean it's not there. Their feedback at this stage is critical.

Exit interviews are an overlooked aspect of the employee experience because that's usually where the most constructive and brutally honest criticism comes out. But rather than seeing the problem as something that can't change because the employee is leaving, the company has an opportunity to invest in those still employed.

In exit interviews, be sure to ask open-ended questions about people's experiences within their roles, such as:

- What worked well during your time here?
- What didn't work well? What would you improve?
- What allowed or didn't allow you to grow in your role?
- Did you feel included here?
- How were you treated here?
- Did you experience harassment, microaggressions, or similar harm while working here? If so, can you share more?

Most importantly, be open to feedback. Use these interviews as a space to listen and collect information, not to defend the company. The role of managers in exit interviews is not to defend themselves or the company but to learn from the employee who is leaving. That employee needs space and a sense of support to share why they're leaving.

If any feedback comes as a surprise, it's a sign that gathering feedback should happen more regularly at the team or company level, perhaps at the same pace as performance reviews. It could also be a sign that you need to check in with current employees about how they're doing and what they're experiencing. That way, you can catch problems sooner and not wait until more employees are driven to leave. If your company doesn't have exit interviews, contact your HR leaders and ask that the practice be put in place, along with a process for taking feedback and turning it into action promptly.

Processing feedback

Once feedback is gathered, determine how to share that feedback and take action. If the feedback stays in an exit interview, the company won't evolve or make the workplace more inclusive. If existing employees are surveyed for feedback and nothing is done with it, their trust may dwindle, and they, too, may eventually leave. Here are a few suggestions for handling feedback:

- Tell the team how feedback will be processed and what they can expect to see.
- Create shared documentation to compile and review feedback and applicable, actionable items.
- Schedule regular debriefs for exit feedback or hold monthly or quarterly meetings, where leaders are invited to learn more and self-assign tasks to improve working conditions.
- Provide constant and regular updates to the rest of the team about your progress.

Changes can only be made if feedback is shared, acted on, and prioritized. Receiving and acting on feedback can become a part of a manager's goals to help retain their existing team. While that will be hard and can feel risky, it's an opportunity to create inclusion.

LEADING THE CHARGE

This chapter is packed with suggestions for leaders, in title or not, to step up and actively demand that more consideration take place for how the workplace operates. It may seem like a lot is out of our control, especially for those not in formal leadership roles, but we can make an impact, even if we don't have an official title. It's about identifying opportunities, doing the research to find solutions that work for the specific areas of improvement, and finding ways to collaborate with colleagues at any level. It's also about communicating how the company and its employees can improve. Anyone can lead that charge.

6 BROADENING THE COMMUNITY

I once worked on a project to help set up a scholarship to increase representation for Black, Indigenous, and Latinx communities in tech. My task was to set up the application form, which captured demographic data—standard, expected information I had seen countless times on other forms. I put the form together and shared the link with the project partners, expecting to maybe hear back about a typo or missing comma.

Instead, one person replied: "Sam, why is 'white' the first listed demographic option?"

"I—I grabbed the list from the US Census," I said, suddenly realizing that I had made two mistakes.

The first mistake was that I relied on a source without considering its implications. I never questioned if it made sense or could cause harm.

The second mistake was that I hadn't considered the context of the form design in the first place. The goal of the scholarship was to create space for marginalized students, but by listing their identities lower on the form, I was perpetuating a harmful status quo.

These two big mistakes could have been harmful to the users, and yet only one partner had noticed them. In the end, alphabetizing the demographic labels was a better—and more intentional—design decision.

By being *intentional*, I mean we pause and think through the goals and implications of what we do or see. There are dozens of opportunities for more intentional care in our communities, even beyond our classrooms and offices. I believe community members have the ability—and the responsibility—to ask questions, demand change, bridge gaps, and educate one another.

Meetup groups, clubs, mentoring spaces, workshops, conferences, and other community places enable designers to continue to learn new skills, broaden perspectives, and flex their growth mindsets. These spaces are not immune to some of the same challenges of the institutions, but they hold a unique opportunity to improve our industry.

GATEKEEPING IN DESIGN CLUBS

Some of design's oldest institutions are professional design clubs and organizations. They hold national events and conferences, which require membership, along with chapter-run local events. A board usually operates them, and some sponsor student- and professional-level design competitions. There are many benefits to joining these organizations, like networking and recognition.

But they aren't immune to the same systemic issues that our schools and workplaces face. Among the designers I know, some love belonging to professional design clubs, but just as many want nothing to do with them. Those who are anti-club cite concerns with:

- high membership fees,
- all-white boards and speaker lineups,
- cliquishness and favoritism among more well-known designers, and
- refusal to hear and address feedback from attendees.

Professional clubs should try to include as many people in the profession as possible. They should make their events welcoming and accept feedback from people who want to attend.

Institutionalized racism in our industry

In 2020, in a post titled, "Dismantling White Supremacy Culture Within AIGA," American Institute of Graphic Arts (AIGA) chapter organizer George Aye wrote:

> The American Institute of Graphic Arts, known as "the professional association for design," dropped their longhand name and went with the acronym AIGA. But with longstanding issues of diversity, equity, and inclusion within the organization, patterns of fragility that perpetuate white supremacy culture and the silencing and erasure of Black voices and labor, I think we should check what the 'A' in AIGA stands for.
>
> Does the 'American' it refers to include this new, emerging definition that reflects where America is heading, or does it mean 'White American' as so many have believed for a while? (https://samk.app/idc/06-01/)

The post ended with a letter of recommendations to AIGA, including advice on governance, operations, strategy, and other areas the organization needs to change.

A month later, Amélie Lamont left their AIGA chapter position (https://samk.app/idc/06-02/), citing concern with the organization, including AIGA's uncredited use of the work and words of designer Antoinette Carroll (https://samk.app/idc/06-03/). Carroll had stepped down from AIGA in 2019 after leading the charge on many diversity and inclusion initiatives, yet still not seeing improvements within the national club and local chapters (https://samk.app/idc/06-04/).

AIGA isn't alone in its lack of equity. Juan Villanueva resigned from the Type Directors Club in 2020, citing racism within the organization:

> I'm resigning because the TDC board doesn't foster a collaborative environment and is not truly open to change. No one

should be accused of trying to break a window to let people into a space designed by white men, and led by white people whose roles throughout my tenure have been to oppress and silence my initiatives such as the Type Crit Crew and the BIPOC fund. (https://samk.app/idc/06-05/)

Girl Develop It (GDI), a nonprofit focused on helping women enter the software industry, also experienced a series of resignations in 2018 after white leaders failed to address racist behaviors and a lack of support for Black members and chapter leaders (https://samk.app/idc/06-06/). Complaints against GDI escalated publicly, ending in an open letter to their board, which GDI disregarded:

> The GDI Board and HQ have yet to appropriately respond to this open letter and have demonstrated that salvaging what is left of the brand is the top priority. There is no coming back from almost a year of inaction and refusal to be held accountable for the harm that was caused within the organization. There is no path forward at this point. GDI must come to an end. (https://samk.app/idc/06-07/)

A common thread among these resignations is a lack of action: all these designers of color and their allies repeatedly tried to provide feedback and guidance to make these clubs more inclusive for the members they purportedly serve; repeatedly, they were ignored. Even when organizations did respond with the intent to improve, it was too little, too late, for many.

These organizations have failed their communities. They've harmed the leaders who have had to resign, the members who have been marginalized, and the community and would-be members who have seen only exclusion.

Enacting change

In 2017, Timothy Bardlavens left AIGA, stating that the organization didn't represent him as a Black designer and that he hadn't seen any interest in change from those in power. Bardlavens was left with many questions:

What is AIGA doing about [inclusion]? How is it pushing from a national level to get the industry more diverse? Are we providing pathways for recent graduates of color, who had to use their university labs to do their design work, get a Mac and subscription to Adobe to continue to build their portfolios while transitioning into professional careers? Are we, from a national level, searching for ways to get funding into schools, especially in lower-income areas, to create graphic design classes, thus introducing more people of color to design sooner? Are we creating messaging for parents of color to show the impact and viability of design as a career? (https://samk.app/idc/06-08/)

These are the types of questions we should be asking every design organization. Examine the professional organizations you're a part of and speak up if you see room for improvement. Here are a few suggestions:

- Write to club boards asking for change or listing instances of discrimination you've seen or heard about. State the harm and be specific about what you expect them to change. Share examples of what is and isn't acceptable or include suggestions.
- At events, ask sponsoring companies to review the event organizers' behaviors. Ask how they vet whom they donate to or how the event aligns with their values. A designer was once announced as a conference speaker representing a church that believed in conversion therapy and punishment for queer kids, making the conference less inclusive. I—and a collection of other individuals and design club chapters—asked the sponsors and other speakers at the conference if they were aware of this and if they aligned with those values. Many sponsors and speakers dropped out of the lineup or withdrew sponsorships once they learned the organizer would not remove the noninclusive speaker.
- Reach out to speakers at events and ask them to demand more inclusivity from the host organization. I've done this when I've noticed colleagues speaking at events hosted by clubs practicing exclusionary behaviors. I first ask if I can share my concerns with them about the event. If they

accept, I will share evidence (links and screenshots) related to my concerns. I don't pressure them to leave the event but let them know what the event is or isn't supporting. Some colleagues have stayed on to speak only if the exclusionary practice against attendees or other speakers is stopped. In contrast, others have removed themselves from the event at their discretion.
- Consider joining the board and becoming a part of the change. However, if you're marginalized in the design world, be careful about ending up in an organization that may not support you, as we've seen from previous examples. If you're privileged in any way, make it your mission to support those marginalized in the organization as members or board members. That said, remember to focus on improving things for everyone and not centering yourself.

If you don't see positive change, cancel your membership(s). Don't participate in organizations you believe are causing harm. Put your money into alternatives (such as scholarships or nonprofits), find inclusive clubs or online communities (such as Queer Design Club or start a networking organization that supports designers and leaders from all backgrounds.

We need to hold our professional clubs accountable if we want them to be accessible to all. While it is the responsibility of organizational leadership to take action, we can demand that action—individually and collectively.

BRIDGING THE GAP

As we saw in Chapter 3, a lot of university curricula are out of date. When I was teaching, Introduction to Graphic Design projects used photocopy machines to manipulate text—something that can now be done quickly with design tools like Photoshop—and Flash was still taught, even after it had effectively died in professional contexts. Students had little access to working designers to show them alternatives.

But while educators may be constrained by their curricula, working designers can help bridge these gaps. You don't have to look far to guide the next generation of designers and make the industry better for everyone.

Giving back in educational spaces

In my early teaching days, I put together an advisory board of designers and developers from my network and local meetups. Advisory boards aren't common in academic departments, but I wanted to get their thoughts on some of the emerging technologies my students needed to know about.

I brought my advisory board members into the classroom in various capacities. Many wanted to get public speaking experience and enjoyed being able to guest lecture in my class or the department. They'd stay longer to answer questions and get to know some students by name. They'd recognize those students at meetups and continue to guide them as they prepared for graduation and their first job search. Others liked to come in for portfolio reviews, sit with students individually, and assess their résumés. This demystified so much of the working world for the students.

It benefited the design professionals, too. Talking about their work with eager audiences rekindled their enthusiasm. They learned from students how difficult the path from school to career had become and how techniques like responsive web design were being taught in the classroom. They also saw firsthand how motivated and talented the students were, and hired many directly after graduation.

If you want to get similarly involved in education spaces, reach out to local schools and see how you might be able to help. Most schools publish contact information for the department chair or office on their website and are delighted to hear from working designers. It benefits them and the department when they can bring in visiting professors, guest lecturers, or mentors for their students.

Once a connection is established, talk about what you can offer, and find out what might benefit the students in the department. Whether it's a technical skill or an industry insight, sharing this information with students and educators is the best way to make an impact.

In the classroom

You don't have to become a professor to get teaching opportunities. Offer to come in as a guest lecturer or speaker. You can share your career experiences or deliver a talk on a topic relevant to the class's curriculum. Or, you can offer to cocreate assignments with professors that reflect more real-world applications.

If teaching isn't for you, you can provide industry feedback to students. Conduct mock interviews so they'll know what to expect when they hit the job market. Offer to review cover letters, résumés, and portfolios, and share what you think employers are looking for.

Mentorship is also an excellent option. Reach out to department chairs or professors to get involved in (or help set up) a mentorship program. You may be able to work with one or two students each semester, answer their questions, provide feedback, and share your expertise. Tell them what you wish you had known. Did you think about how you'd negotiate your salary as a new designer? Did you wonder about benefits and company culture? Share whatever professional knowledge you have.

In administration

You can also offer to review the design curriculum and share recommendations with the faculty. Universities have slow timelines for updates, usually going through multiple rounds of governmental and board reviews to ensure they'll still meet certification requirements. Getting involved in this process could help bring the curriculum up to date, better equipping students when they enter the field.

Professional opportunities

Invite students to see your work. Bring in or link to actual deliverables of a recent project that you're able to share while visiting their classrooms. Or, bring the students to your workplace so they can see what a working design studio looks like.

Offer internships to students. Many schools require internships; work with hosting companies to ensure students are credited and paid at least minimum wage.

Finally, try to connect students with other professionals in your network. The more professionals a student can meet, the more perspectives they'll hear, and the more smoothly they can transition into the industry. Try to pair students with professionals with similar interests, backgrounds, or career goals.

Unlearning and relearning at work

The previous examples have been about connecting with students, but they aren't the only way to make an impact. Too often we expect students to adapt to the industry, without considering how we should adapt to them. In our workplaces, we can try to change outdated conversations that may have taken hold during our own educations:

- **Model broader influences.** There's a good chance your design team needs to learn more than Eurocentric design history. Seek out project inspiration from global design movements. Resist naming the same white male designers over and over. Use team lunch 'n' learn sessions to study work by Black and Indigenous designers or design history from other parts of the world.
- **Use your voice in local events.** Designers in the industry have more clout than design students regarding events around town. Push local design clubs to host events on global design history and host speakers with marginalized identities. If you are an event organizer, the rest of this chapter offers greater detail on how to run inclusive events.
- **Revisit (paid!) internships and other parts of the workplace.** Talk to your hiring managers about internships and other opportunities to bring students into the workplace. Internships get a lousy reputation because employers assume that interns require a lot of handholding—which isn't necessarily true. Find ways in which having an intern could be beneficial to your organization. Consider how accessible your office is, how meetings can be made to include everyone,

and what documentation you can support for onboarding new employees. These steps will benefit everyone, not just students and self-taught designers.

A gap exists between those learning outside of the workplace and those learning within it. Suppose we engage in being transparent and supportive of one another, and we work collectively to push for more inclusive topics, history, and processes within events. In that case, we can start to bridge the gap between education and other spaces where conversations about design exist.

DEMANDING BETTER EVENTS

Conferences, workshops, and events in the design and tech industries aren't always the most inclusive spaces. While some events are models of inclusivity, others create inaccessible or unsafe spaces, exclude historically marginalized groups from speaking opportunities, or tokenize speakers by asking them to speak solely on diversity.

Whenever I'm considering attending or speaking at a conference, I ask many questions of organizers and former speakers in my network to ensure the event is inclusive. Over the years, I've identified several behaviors from event organizers that act as red flags:

- They talk about inclusivity, but can't share specific steps they've taken (beyond insisting that their attendees "come from diverse backgrounds").
- They copy and paste codes of conduct from other conferences. I once read the code of conduct for an in-person conference run by a local chapter of a larger organization and realized it was written for online events—the conference organizer hadn't read it before sharing it with me. This let me know they hadn't considered creating a safe and inclusive environment for those attending.
- They exclude marginalized representation when discussing inclusion. I was once invited to speak on a conference panel

about diversity, where it turned out I was the only woman of color, and there was no disability representation.
- They underpay (or lie about paying) their speakers, or they pay speakers with marginalized identities less than speakers with dominant identities. (You may only find this out during or after the event.)
- They invite or allow known bad actors to speak or attend, or they don't handle harassment complaints in ways that protect those who are harassed.

Not all hope is lost, however. Many conferences work hard to make their events inclusive, showcase representative speakers, and publicly post their codes of conduct and enforcement guidelines. Whether you're attending, speaking, or organizing, there are plenty of opportunities to improve these spaces and make them inclusive, inviting, and safe for everyone.

Inclusion at every level

There are many effective ways conference and event organizers can make their events more inclusive—many surprisingly easy and -inexpensive. Women Talk Design (https://samk.app/idc/06-09/), an online resource for women and nonbinary speakers, recommends taking a holistic approach to inclusivity at conferences. In an article for Women Talk Design (https://samk.app/idc/06-10/), Angela B. Brown, head of events at The Linux Foundation, suggested that events provide inclusive options like:

- Unisex T-shirts in a wide range of sizes
- Gender-neutral restrooms
- Quiet rooms for introverted or neurodivergent attendees
- Private childcare and nursing rooms
- Disability accommodations at the venue

Confab (https://samk.app/idc/06-11/), an annual content strategy conference, includes a "quiet room" at the event, where people can recharge without social interaction, and a "noisy room" for those who want to socialize and engage directly with other attendees. They provide buttons for attendees that indicate how they want to be approached (using colors like red, yellow, and green); a lactation room with a refrigerator; a locking door for privacy; and reading materials for nursing parents. They also check for dietary restrictions.

Many conference organizers can learn from Clarity, a design systems conference run by Jina Anne, a true force in the design community (https://samk.app/idc/06-12/). Clarity is one of the most inclusive conferences I've ever attended or spoken at, in person or online, during the pandemic. The Clarity team has done a tremendous amount of work to make sure the event creates a safe environment for all by considering:

- **Accessible spaces.** Clarity tries to pick venues with gender-neutral bathrooms and wheelchair access and seating. Videos of talks are eventually released with captioning, and during the virtual event in 2020, a live captioner transcribed all twelve talks. When hosting online events, Clarity uses web meeting applications with built-in accessibility tools like captioning and transcribing.
- **How attendees want to interact.** Attendees can customize conference badges with their pronouns and their preferences about being photographed or recorded; this way, they can be upfront about their needs without having to explain themselves repeatedly.
- **Clear communication.** Jina Anne shares everything about Clarity's commitment to inclusivity on its Diversity and Inclusion page (https://samk.app/idc/06-13/), letting speakers and attendees know about the conference's values in advance. The page provides as much detail as possible, including when the audience can expect updates or more information.

When conferences center the users—the attendees and speakers—the way we'd expect designers to do, it leaves a memorable impression, creates trust, and encourages repeat attendance. Many conferences—and workplaces and classrooms, too—can learn from Clarity's example.

Mindful attendance

As an event attendee, it's important to be mindful of how you share space with others—even if it's just for a short period, events *are* a shared space, a community. In a community, there are unspoken rules for how we interact and conduct ourselves; for instance, we know to keep quiet during a talk and clap at the end of it. But mindful attendance is also about awareness of the community and how we can improve everyone's experience, as well as anticipating who might need something within the space, or who might be missing from it.

As with a lot of my other suggestions, I ask that you think critically and do your research. When considering whether to attend a conference, make sure it's inclusive by asking questions like:

- How does the conference make tickets available to historically marginalized communities? Conferences invested in access and inclusion often offer free tickets or scholarships to people from marginalized backgrounds.
- How does the conference ensure the safety of speakers, attendees, sponsors, vendors, venue staff, and volunteers during the event? Do they have a code of conduct? How is it enforced? (We'll talk more about codes of conduct later in this chapter.)
- What accommodations are provided? For example, does the conference have gender-neutral restrooms, wheelchair access, and ramps for the stage and entry into the venue? Are talks transcribed? Are there captions, or is sign language interpretation provided? Is there a private nursing room available? Is there childcare? Knowing what accommodations are provided, even if you don't personally need them, can tell you how inclusive the event is trying to be. Confab's philosophy of "treating each attendee as a VIP" means they consider everything from dietary needs to religious practices when scheduling their conference each year—and they have answers for anyone who asks about accommodations. When organizers don't have answers ready, it's a sign that there may be other barriers to entry into the event.

- How is this event bringing in an inclusive lineup of speakers? How are the conference organizers actively seeking out speakers from marginalized communities? Are speakers asked to speak only about their experience of being marginalized, or about the conference's subject matter and area of expertise?
- Are all speakers paid? If so, how much, and are they equally compensated? Like the transparent salary conversation, the more we talk openly about speaking fees, the more easily we can identify when speakers are being unfairly compensated due to race or gender.
- Can speakers or attendees opt out of being photographed? Since conferences are in public places and usually documented with photos or video, it's an excellent idea to allow attendees and speakers to opt out of being photographed. They don't need to share their reasoning, but there are many valid reasons: they may be worried about their safety inside or outside of the event, or they may be undocumented immigrants, or they may simply prefer not to be photographed. The same rules should apply to remote events: never screenshot anyone who has not opted into being recorded. Privacy is a part of safety.

This isn't an exhaustive list, and not all of this applies to every individual or event. Think about what matters to you. If a conference website doesn't clarify its values, policies, and accommodations, reach out to the organizers and ask for what you need through email, social media, or some other written form. That way there is a record, and it will be easy for them to send documentation if they have it.

Intentional speaker lineups

Inclusive speaker lineups are not challenging to put together. But when conference organizers seek out only popular or well-known speakers, they end up with conferences that cater to celebrity and privilege, repeating the same event over and over.

If you're organizing an event, consider how empowering it can be for attendees to hear from speakers representing their experiences and identities. Seeing speakers like themselves on stage may lead to thoughts like, "Next year, I want to do that," or "I didn't think I could do that, but now I can imagine it." Think about the story your conference tells when you have intentional speaker lineups; the event experience will be enriched for everyone involved.

Diversify your lineups

Clarity's conference does not accept formal proposals to speak, because those usually come from well-known speakers who typically have more experience or resources available to them than others:

> About 90% of the people who reach out [to speak] are cis-gender straight white men. If I only relied on these requests, this would not make for a very diverse or inclusive event. [...] I made the decision at the very beginning that ***I will personally curate a diverse range of perspectives***—not just for race and gender—but also in subject matter expertise, experience levels (I love inviting people who are brand new to speaking), healthy differences in opinions, and many other factors. (https://samk.app/idc/06-14/)

Jina Anne's intentionality around speaker curation helps ensure that Clarity is welcoming to new speakers. Similarly, to support new and potential speakers, Confab offers advice on how to submit a conference talk proposal on its site (https://samk.app/idc/06-15/). This creates a more inviting event by demystifying the speaking and proposal process for less experienced speakers.

Be intentional about creating your speaker lineups:

- Establish (and stick to) goals for panels and lineups, such as "no all-male or all-white panels" or "more than half of all speakers self-identify with a marginalized group."
- Ask attendees, former speakers, and colleagues whom they'd love to hear from—especially those who haven't spoken at a conference before. Share incomplete speaker lists when

inviting potential speakers. They may have suggestions (and will enjoy knowing whom they will be backstage with).
- Invite potential speakers directly, and provide all the details about the event (including payments) with your invitation. Expanding methods for outreach to speakers beyond calls or requests for proposals (CFPs or RFPs) can lead to a more diverse lineup.
- Pay all speakers the same speaking fee and cover their accommodations, including a free ticket to the event. If they push back on the same fee for all, explain that you do this to be fair to all speakers. Clarity does this well, offering alternatives to the fees such as honorariums or donations that cover scholarship tickets for marginalized attendees.
- Be transparent about the steps you've taken toward inclusivity, and how you still want to improve. Let people know you're looking to have a more inclusive lineup. Use social media to get the word out and share how potential speakers can connect with you.

As in other topics we've covered, reframing how we think about experiences and making key pivots in our actions can change what happens in a community. Transparency and intentionality can help build trust between a conference and its attendees.

Marginalized speakers on marginalization

In recent years, more speakers and events have been presenting talks about diversity and inclusion. While this seems like a good thing on the surface, there are complex dynamics in play.

People from marginalized groups are often asked to speak only about their differences or identities, or they're made to feel like they're filling a diversity quota. This tokenism can be harmful; not only does it hamper the speaker's career goals, pigeonholing them into that topic, but it also signals to attendees that people with marginalized identities can only speak about their marginalization. The speaker and the audience both miss out on what they have to offer about the design process, technique, craft, theory, strategy, and aesthetics. Break barriers by getting marginalized speakers to present on a range of topics—ask them what they'd like to present.

If you've been marginalized as a designer, you may want to share that experience so others can learn from it. Only do this if you have the emotional bandwidth to do so—and know it's not your responsibility always to play that role. It takes a lot of energy to discuss your experiences (some of which might be traumatic) under a spotlight. I find myself going back and forth between talking (and writing, ha!) about marginalization and not. Some days it's tough and heartbreaking; other days I feel a need to speak up. That pendulum swing is normal, and setting whatever boundaries you need is okay. As they say on airplanes, put your own oxygen mask on first before you help anyone else.

It's also okay to expect or ask more from other speakers. No matter the topic, speakers can influence how an event is run because they're in a position of privilege. They have a direct line to event organizers to share concerns, and can be in spaces attendees don't usually get to see or experience.

The power of speaking

Designers and developers have increasingly been using inclusion riders to better wield power as speakers and enhance inclusion in the industry. Because speakers provide the core content of any event, they are responsible for demanding better conditions for everyone at a conference and ensuring that people from marginalized identities can be included.

Inclusion riders were created for the film and television industry by professor Stacy L. Smith, attorney Kalpana Kotagal, and actor and producer Fanshen Cox:

> [A]n inclusion rider is legal language that actors can embed in their movie contracts, which guarantees the project achieves a certain level of diversity both on-screen and off. (https://samk. app/idc/06-16/)

Inclusion riders can include requirements about the safety of an event, accommodations, speaking topics, and more. Designer and speaker Tatiana Mac posts their inclusion rider publicly on

GitHub (https://samk.app/idc/06-17/). It lists detailed requirements for in-person and remote events, travel and lodging, speaker representation, codes of conduct, scholarship programs for attendees, and speaking fees.

Speakers can also use a speaking invitation as an opportunity to push organizers for inclusive change. One of the first questions designer Ethan Marcotte asks event organizers about (via email, so it's documented) is speaker compensation—not only his own, but the compensation for other speakers as well, to ensure payment is fair:

> *"How do you plan to compensate your speakers?" [is] a logistical question. But in the spirit of full disclosure, this is also the question [I've] revised the most.*
>
> *Historically, I've liked being up front about my terms. [...] If their budget didn't line up with my fee, for example, then we could save a few rounds of back-and-forth over email.*
>
> *[But this wording] lets me start the conversation around how the organizer thinks about compensation for all their speakers, not just me.* (https://samk.app/idc/06-18/)

It's also common to ask organizers about accommodations, codes of conduct and enforcement, and audiences. The goal of asking questions and using inclusion riders is to get event organizers (like workplaces) to understand and own the responsibility of building toward inclusion instead of leaving it up to chance.

Codes of conduct

I was once part of an Austin-based design and developer Slack group where several male designers were harassing women. When several of us asked the (white, male) organizer of the group to create and enforce a code of conduct, he laughed us off, saying that he didn't want to "be the bad guy" and that the group members "are adults"—except that neither he nor the harassers were acting like it.

As a result, most of the women and their male allies left the Slack group, which eventually fell apart. The organizer had promoted male dominance and ignored harassment—all while worrying a code of conduct would make him look bad.

Codes of conduct are documented guidelines for community behavior. We see them in places like employee handbooks, conference websites, online design communities like Dribbble, and chat groups like Slack. Every community should have a code of conduct that is publicly accessible. They should be proactive and not reactive, and they should exist well before a community event (or incident) takes place.

AlterConf organizer Ashe Dryden has spent a significant amount of time documenting and evangelizing why codes of conduct are necessary:

> *The people most affected by harassing or assaulting behavior tend to be in the minority and are less likely to be visible. As high-profile members of our communities, setting the tone for the event up front is important. Having visible people of authority advocate for a safe space for them goes a long way.* (https://samk.app/idc/06-19/)

Good codes of conduct set clear boundaries and rules for engagement, with the aim of making the community safe for everyone involved. If you're hosting meetups online or in person, you should write a code of conduct, share it before the event, and send out periodic reminders to community members.

Specificity

Organizers are responsible for ensuring events run well, and part of that is creating a custom code of conduct for the event. Too many organizers in our industry simply copy and paste generic codes of conduct, which is just as harmful as not having one. On CSS-Tricks, Chris Coyier noted that most conferences he had come across were linking to confcodeofconduct.com, a template for an open-source code of conduct, rather than writing and enforcing their own:

> *The primary concern about linking directly to someone else's code of conduct or copy and pasting it to a new page verba-*

tim is that **there is nothing about what to do in case of problems.** So, should a conduct incident occur, there is no documented information for what people should do in that event. Without actionable follow-through, a code of conduct is close to meaningless. It's soul-less placating. (https://samk.app/idc/06-20/)

It's also important to consider the specific audiences and settings for the event. Clarity's code of conduct covers interactions in their online Slack channel, online events associated with the conference, and in-person events. It's also clear *whom* the code applies to: not just speakers and attendees, but also partners, sponsors, volunteers, vendors, venue staff, conference staff, and others. The code specifies how to get support when needed, what happens when a report is filed, and how to request accommodations.

In 2018, Austin Design Week organizers asked me to join their board. Because we had multiple events running simultaneously in various locations, I voiced to the board that we should improve the code of conduct so it could scale and align with the organization's values and make sure it was enforceable. We made many changes that year, including:

- Adding guidelines around online conduct to pair with our updated in-person guidelines
- Labeling board members as moderators for the Slack channel to ensure attendees knew whom they could reach out to if they experienced any misconduct
- Providing multiple online and in-person communication options for contacting board members and volunteers
- Adding a form to report incidents, along with explanations of the reporting processes
- Repeatedly sharing the code of conduct online and talking about enforcement with attendees at every event
- Training volunteers on how to handle misconduct

These changes allowed more people to be responsible for the safety of all events across the city (https://samk.app/idc/06-21/).

Enforcement and training

Writing a code of conduct, while necessary, isn't enough. Having guidelines and procedures for enforcement is crucial to ensuring that the code of conduct works.

When attendees know exactly how an incident will be dealt with, they're more confident that organizers have safety in mind, and will be more likely to report problems when they happen. If the process is mysterious or unclear, attendees may feel that reporting won't increase their safety or that it won't be handled appropriately. They'll be more likely to leave the event or community, and bad actors will be more likely to continue harmful behavior.

Queer Design Club explicitly states the consequences for breaking their code of conduct:

> *[A] temporary ban or permanent expulsion from the community without warning (and without refund in the case of a paid event). Queer Design Club admins may choose to identify violators of our policies against abuse and harassment as a harasser to other Queer Design Club members or the general public. We may also choose to share that harassing or abusive content within or outside of the Slack channel if we feel it is best for the safety of the community.* (https://samk.app/idc/06-22/)

They also outline how violations will be addressed, including violations committed by organizers. This creates space for safety and accountability.

Once the consequences and procedures are clear, train your staff and volunteers to handle reports of misconduct. Steve Fisher, founder of the Design and Content Conference, wrote:

> *Conference staff and volunteers must be trained and prepared. It's harmful to have a code of conduct that event organizers aren't prepared to enforce. It creates a false sense of safety and action. It's a broken trust that will hurt people.* (https://samk.app/idc/06-23/)

You can also map out common misconduct scenarios and discuss how to address them as a group. Practicing these scenarios can build confidence to step in when problems arise.

Attending events

In 2015, designer Rachel Nabors was invited to speak at Jared Spool's UX Immersion conference. But when they requested a code of conduct, they were told Spool held to his publicly anti-code-of-conduct stance, stating that professional events are safe enough on their own and that having one would not be effective. Nabors could not change Spool's mind and would not speak without a code of conduct, so Spool retracted the invitation. Nabors then shared the experience on their blog, saying:

> Of course I worry that writing this might mean getting invited to fewer UX conferences. But it was more important to explain, in minute detail, all the issues with this line of thought, both for the benefit of other organizers and to assure other women that no, you are not irrational for expecting these things from the events you attend.
>
> I spoke with many women (and men) about this, even ones who have spoken at Jared's events in the past, and their advice was the same: if you feel strongly about something, put your foot down. Demand the change you want to see. It's the only way to raise industry standards. (https://samk.app/idc/06-24/)

Organizers who object to codes of conduct misunderstand them. They're unwilling to take responsibility for their events, expecting large groups of people to manage themselves or assuming that nothing can go wrong. This creates opportunities for chaos and offers no protection or recourse for victims of harassment.

When considering joining a group or event, ask the organizers for their code of conduct and how they plan to enforce it. If they don't have one, request that they write one specific to the event, including reporting procedures and plans for enforcement (https://samk.app/idc/06-25/). Don't speak at or attend conferences that lack a code of conduct or can't provide answers to your concerns.

Codes of conduct don't guarantee safety; they put a plan in place for when things go wrong and signal to community members that safety and inclusion are a priority.

IT TAKES A COMMUNITY TO MAKE A COMMUNITY

What excites me about writing this chapter is how many possibilities we each have to impact and improve the design industry. As individuals, groups, or events, as speakers or attendees, we can each choose actionable items that will make our communities more inclusive and inspiring for everyone.

In the first chapter, I wrote about doing the work. This is some of that work. And if everyone takes even one or two of these small steps toward inclusivity, I believe we'll see change. That's collaboration, and that's community.

CONCLUSION

We've journeyed far and explored a lot together, and I recognize your investment in creating a better, more inclusive community for everyone in it. Including you.

I hope that whatever background or identity you bring to reading this book, you've found some support and encouragement to meet the situations you might be dealing with. I hope you feel a bit more equipped to make the design community better. I hope you have gained new perspectives on how inclusion and exclusion manifest for people with different identities, and have learned new ways to use privilege and power to help others. I hope you feel inspired to advocate for others' needs and rights, even if doing so doesn't directly serve you. I hope you know you are not alone, and you deserve to have good experiences in design spaces.

Because you chose to read this book, I believe you want to make our industry better for everyone in it, today and into the future. You've undoubtedly realized how challenging this work can be, but also that it simply can't wait any longer. Yes, it takes time to learn new things, unlearn engrained habits, imagine other perspectives, and grow. And sometimes we don't act because we're afraid we'll make mistakes. Fear and inaction are what built exclusive spaces so sturdily in the first place. Don't let fear stop you from stepping up and moving forward.

Start with just one action or contribution, and then try another. Keep going.

ACKNOWLEDGMENTS

I'm indebted to Katel LeDû, Lisa Maria Marquis, Sally Kerrigan, Susan Bond, and Kumari Pacheco for offering support every time imposter syndrome reared its three heads. I treasure your advice: reminding me when it was time to *write* and when it was time to *edit*. You helped me work through the many fears I have about writing. Thank you for seeing what I didn't know was possible for me yet. I can't thank you enough for that, and for sharing the party Gritty emoji in Slack. And thank you to the whole ABA team for the work you did to make this book happen.

Becoming an educator, then a manager, and then writing a book would have been impossible without Jeff Davis and Bill Meek. You're the first two mentors I had, and your patience and belief in me helped me write my master's thesis as well as this book. So much of what I know about academia and advocating for others I learned from you.

Thank you to mentors, coaches, and friends who always pushed for my growth: Jeffrey Zeldman, Greg Storey, Ethan Marcotte, and Gene Crawford. You've supported me and my students over the years and made it possible for so many more designers to be in this industry. Thanks to Peter Barth and Kyle Fiedler for creating opportunities for me to learn and grow as a manager and trusting me to do so. Kyle, thank you for every bit of support you provided for this project. Sara Wachter-Boettcher, Mia Blume, April DiMartino-Neufeld, Tara Chivukula, and my Within (cactus) coaching circles: thank you for being incredible coaches and listeners along the way, and for inspiring in me a deep sense of belonging.

Daniel and Alyse, Lachlan Hardy and Elle Meredith, Eric Bailey, and Melanie Richards: many thanks to each of you for creating space for me to talk through so many parts of this book. Thank you Eric, specifically, for being a helpful sounding board, an inspiring writer, and for teaching me that "computers were a mistake." Enormous gratitude to the many people I interviewed or surveyed, including Tenessa Gemelke, Omari Souza, Jennifer Clark, and more than twenty design and development managers who shared their experiences.

Vanessa Crook, Hanson Little, Graci Willis, Adekunle Oduye, Cassidy Browning, Moyo Oyelola, Tiffany Stewart, and many other friends: thank you for being supporters in this writing process and for the meaningful conversations about identity, race, immigration, and marginalization over the years.

Finally, I'm grateful to my parents, Deepak and Surina, who took their young girls on a first-generation immigrant family adventure, giving us the most wonderful life in a beautiful place full of culture: Curaçao. My passion for building communities is 100 percent genetic and something I've seen you both model my whole life. My sister, Deepina, and my brother-in-law Ryan: thank you for breakfast tacos and chats about inclusive workplaces and management too numerous to count. I love you. Thank you.

RESOURCES

One of the best ways to begin making design communities more inclusive is to learn from folks who have been actively doing this work already, and follow their lead.

Equity and inclusion fundamentals

- Self-Defined by Tatiana Mac and other open-source contributors has greatly impacted our understanding of inclusive language in design and the tech industry (https://samk.app/idc/07-01/).
- The Conscious Kid has great resources for how to talk to children about disrupting racism, inequity, and bias (https://samk.app/idc/07-02/).
- Author and educator Blair Imani's "Smarter in Seconds" YouTube channel has short but detailed videos on terms, ideas, and history relating to inclusion (https://samk.app/idc/07-03/).
- Various inclusion-focused organizations use different descriptors for identity. The USC's School of Social Work Diversity Toolkit breaks identities into two groupings—target identities and nontarget identities—and explains the intersections between them (https://samk.app/idc/07-04/).
- Rachel Cargle's article in *Harper's Bazaar* about toxic white feminism tackles how people of color are harmed by both good and bad intentions (https://samk.app/idc/07-05/).
- Sylvia Walby writes on the facets of patriarchy (https://samk.app/idc/07-06/).
- *Beyond the Gender Binary*, by poet, artist, and activist Alok Vaid Menon, is an excellent book for any age (https://samk.app/idc/07-07/).
- My personal collection of writing, videos, and resources (https://samk.app/idc/07-08/).

Antiracism and allyship

- Anti-Racism Daily (ARD) releases daily stories, tips, and history lessons about being antiracist (https://samk.app/idc/07-09/).
- Right To Be is an organization focusing on training people about harassment and prevention. One of my favorite courses is "Bystander Intervention Training" (https://samk.app/idc/07-10/).
- Kim Crayton's "Being Antiracist" is a challenging course for many but it shines a light on engrained and less visible elements of racism, particularly from white perspectives (https://samk.app/idc/07-11/).
- Change Catalyst's guide to allyship, "The State of Allyship Report: The Key to Workplace Inclusion," is an important and well-rounded resource (https://samk.app/idc/07-12/).
- Resmaa Menakem uses Somatic Abolitionism to illustrate how history, race, and trauma live in our bodies and souls—and how we can heal and become antiracist (https://samk.app/idc/07-13/).
- An expansion of the groundbreaking work by Nikole Hannah-Jones, *The 1619 Project: A New Origin Story*, contextualizes "the systems of race and caste that still define so much of American life today." (https://samk.app/idc/07-14/).

Inclusion in design

- *Technically Wrong*, Sara Wachter-Boettcher (https://samk.app/idc/07-15/).
- *Accessibility for Everyone*, Laura Kalbag (https://samk.app/idc/07-16/).
- *Design for Safety*, Eva PenzeyMoog (https://samk.app/idc/07-17/).
- "Diversity and Design: How We Can Shape a More Inclusive Industry?", Bronwen Rees (https://samk.app/idc/07-18/).
- "Spotlight on Gender Diversity in Design," Ken Kongkatong (https://samk.app/idc/07-19/).
- "How to Infuse Diversity and Inclusion in Design," an interview with Fabricio Teixeira (https://samk.app/idc/07-20/).

- "Double or Nothing: Can Designers Erase the Gender Pay Gap?", Lilly Smith (https://samk.app/idc/07-21/).
- The Inclusion and Diversity Compendium for Designers: Collective List of Inclusion Resources (https://samk.app/idc/07-22/).

Nontraditional design curricula

- BIPOC Design History provides invaluable online lectures focused on design by the BIPOC community (https://samk.app/idc/07-23/).
- The Accessibility (a11y) Project has a list of resources to help improve accessibility on the web, and in tech and product design communities (https://samk.app/idc/07-24/).
- Airbnb's Another Lens project explains how to use a growth mindset to address bias as a designer (https://samk.app/idc/07-25/).
- A Microsoft inclusive design guide helps designers adopt more inclusive practices in their craft (https://samk.app/idc/07-26/).
- *Queer X Design*, Andy Campbell (https://samk.app/idc/07-27/).
- *Cross-Cultural Design*, Senongo Akpem (https://samk.app/idc/07-28/).
- *This is What I Know About Art*, Kimberly Drew (https://samk.app/idc/07-29/).

Inclusive hiring

- Lara Hogan created a pre-brief template to start the interview process (https://samk.app/idc/07-30/).
- Software companies Greenhouse (https://samk.app/idc/07-31/) and Workable (https://samk.app/idc/07-32/) have built inclusive options into their products.
- The Employer Assistance and Resource Network on Disability Inclusion (EARN) provides resources for recruiting, hiring, and retaining workers with disabilities (https://samk.app/idc/07-33/).

- The *New York Time*'s Wirecutter has a list of the best live transcription services (https://samk.app/idc/07-34/).
- Job boards, resources, and candidate profiles specifically for marginalized groups:
    - Project Include (https://samk.app/idc/07-35/).
    - Circa (https://samk.app/idc/07-36/).
    - People of Color In Tech (https://samk.app/idc/07-37/).
    - AbilityJOBS (https://samk.app/idc/07-38/).
    - Black Tech Pipeline (https://samk.app/idc/07-39/).
    - Career Contessa (https://samk.app/idc/07-40/).
- Writing inclusive job postings:
    - "How to Write an Inclusive Job Post" from PDX Women In Tech (https://samk.app/idc/07-41/).
    - "How to Write Effective and Inclusive Job Descriptions," Elizabeth Black (https://samk.app/idc/07-42/).
    - "5 Must-Do's for Writing Inclusive Job Descriptions," Maxwell Huppert (https://samk.app/idc/07-43/).
    - Gender Decoder provides scoring for use of inclusive language in job postings (https://samk.app/idc/07-44/).

Inclusive work culture guides

- Bookshop.org's reading list for inclusive leaders (https://samk.app/idc/07-45/).
- Inclusion@Work, a framework for Disability Inclusion, from the Employer Assistance and Resource Network (EARN) (https://samk.app/idc/07-46/).
- "Women of Color Get Asked to Do More 'Office Housework.' Here's How They Can Say No," Ruchika Tulshyan (https://samk.app/idc/07-47/).
- "For Women and Minorities to Get Ahead, Managers Must Assign Work Fairly," Joan C. Williams and Marina Multhaup (https://samk.app/idc/07-48/).
- *The Empathetic Workplace*, Katharine Manning (https://samk.app/idc/07-49/).
- Advice and support for diversity, equity, and inclusion at work from Lily Zheng (https://samk.app/idc/07-50/, https://samk.app/idc/07-51/).

- A guide to inclusive meetings from Elaina Natario (https://samk.app/idc/07-52/, PDF).
- Tips for inclusive remote teams from resource diversity consultant Paradigm (https://samk.app/idc/07-53/, signup required).

Diversity initiatives
- The Collective (https://samk.app/idc/07-54/).
- NOVA (https://samk.app/idc/07-55/).
- Paradigm (https://samk.app/idc/07-56/).

REFERENCES

Shortened URLs are numbered sequentially; the related long URLs are listed below for reference.

Chapter 1

01-01 https://chicagounbound.uchicago.edu/cgi/viewcontent.cgi?article=1052&context=uclf

01-02 https://www.law.columbia.edu/news/archive/kimberle-crenshaw-intersectionality-more-two-decades-later

01-03 https://www.payscale.com/research-and-insights/gender-pay-gap/

01-04 https://www.thirdway.org/report/the-fatherhood-bonus-and-the-motherhood-penalty-parenthood-and-the-gender-gap-in-pay

01-05 https://www.bbc.com/news/magazine-29644591

01-06 https://www.hachettebookgroup.com/titles/ijeoma-oluo/so-you-want-to-talk-about-race/9781580058827/

01-07 https://time.com/6072750/kardashians-blackfishing-appropriation/

01-08 https://variety.com/2017/film/news/scarlett-johansson-ghost-in-the-shell-whitewashing-1202020230/

01-09 https://www.npr.org/sections/codeswitch/2020/01/17/406246770/how-namaste-flew-away-from-us

01-10 https://www.theverge.com/22924327/mexico-cultural-appropriation-law-indigenous-and-afro-mexican-communities

01-11 https://www.npr.org/sections/codeswitch/2013/10/21/239081586/the-racial-history-of-the-grandfather-clause

01-12 https://time.com/6117685/book-bans-school-libraries/

01-13 https://www.texastribune.org/2021/12/20/texas-library-books/

01-14 https://inclusion.msu.edu/_assets/documents/bic/BIC-Tips6-Microaggressions-FINAL-Accessible.pdf

01-15 https://www.washingtonpost.com/magazine/interactive/2022/assimilation-chinese-names-asian-racism/

01-16 https://www.chron.com/news/houston-texas/article/Texas-lawmaker-suggests-Asians-adopt-easier-names-1550512.php

01-17 https://medium.com/afrosapiophile/victims-of-racial-profiling-are-in-fight-or-flight-mode-in-public-spaces-and-it-is-exhausting-8b387ff76238

01-18 https://southtampacounselor.com/blog/2021/2/5/understanding-fight-flight-freeze-and-the-fawn-trauma-response

01-19 https://medium.com/facebook-design/navigating-whiteness-part-1-cbeceb86cbad

01-20 https://hbr.org/2019/11/the-costs-of-codeswitching

01-21 https://shegeeksout.com/understanding-intent-vs-impact/

01-22 https://guidetoallyship.com/

01-23 https://www.forbes.com/sites/alexreimer/2020/12/08/nfl-keeps-trying-to-have-it-both-ways-with-colin-kaepernick/?sh=73babf002088

01-24 https://interactioninstitute.org/illustrating-equality-vs-equity/

01-25 http://interactioninstitute.org/

01-26 https://www.collegeart.org/pdf/diversity/white-privilege-and-male-privilege.pdf

01-27 https://twitter.com/mariejbeech/status/1376907398691053577

01-28 https://www.instagram.com/p/B7ybVdXlCUV/

Chapter 2

02-01 https://www.brainpickings.org/2014/01/29/carol-dweck-mindset/

02-02 https://www.mindtools.com/pages/article/learning-zone-model.htm

02-03 https://abookapart.com/products/design-for-cognitive-bias

02-04 https://implicit.harvard.edu/implicit/iatdetails.html

02-05 https://implicit.harvard.edu/implicit/faqs.html#faq12

02-06 https://bookshop.org/books/teaching-to-transgress-education-as-the-practice-of-freedom/9780415908085

02-07 https://www.nytimes.com/2019/01/27/world/europe/metoo-backlash-gender-equality-davos-men.html

02-08 https://qz.com/1099083/analysis-of-141-hours-of-cable-news-reveals-how-mass-killers-are-really-portrayed/

02-09 https://www.sentencingproject.org/publications/un-report-on-racial-disparities/

02-10 https://www.ccl.org/articles/leading-effectively-articles/coaching-others-use-active-listening-skills/

02-11 https://iheartmob.org/pages/bystander-intervention-online

Chapter 3

03-01 https://youtu.be/zQUuHFKP-9s?t=196
03-02 https://www.theguardian.com/artanddesign/2017/mar/28/how-we-made-font-comic-sans-typography
03-03 https://theestablishment.co/hating-comic-sans-is-ableist-bc4a4de87093/#.vukjzcz1b
03-04 https://download.microsoft.com/download/b/0/d/b0d4bf87-09ce-4417-8f28-d60703d672ed/inclusive_toolkit_manual_final.pdf
03-05 https://twitter.com/beep/status/228126115363958787
03-06 https://www.penguinrandomhouse.com/books/612188/this-is-what-i-know-about-art-by-kimberly-drew-illustrated-by-ashley-lukashevsky/
03-07 https://flatironschool.com/courses/product-design-bootcamp/
03-08 https://www.britannica.com/biography/Ada-Lovelace
03-09 https://www.advertisingweek360.com/design-evolution-air-india-maharajah
03-10 https://www.thenationalnews.com/arts/india-s-most-famous-ad-campaign-how-the-amul-butter-girl-has-been-churning-up-debate-for-50-years-1.906660
03-11 https://papress.com/products/w-e-b-du-boiss-data-portraits-visualizing-black-america
03-12 https://queerdesign.club/
03-13 https://www.instagram.com/p/CZPEenuP0ks/
03-14 https://www.guerrillagirls.com/naked-through-the-ages
03-15 https://www.finearts.txst.edu/Art/academics/undergraduate/communication-design/comdes-resources.html
03-16 https://bookshop.org/books/teaching-to-transgress-education-as-the-practice-of-freedom/9780415908085
03-17 https://docs.google.com/document/d/196gKKZPYzl-t6IH39-J8FURK1qXfSNymhlxvl37SnrfU/edit
03-18 https://web.archive.org/web/20210619031301/https://designcreativetech.utexas.edu/response-student-demands-sdct-leadership
03-19 https://stateofblackdesign.com

Chapter 4

04-01 https://techcrunch.com/2015/11/03/twitter-engineering-manager-leslie-miley-leaves-company-because-of-diversity-issues/

04-02 https://www.bloomberg.com/news/articles/2021-06-03/snowflake-ceo-says-worker-merit-should-outweigh-diversity-goals

04-03 https://medium.com/fearless-futures/the-lowering-the-bar-delusion-d17126d4caad

04-04 https://www.wsj.com/articles/the-dangers-of-hiring-for-cultural-fit-11569231000

04-05 https://www.mckinsey.com/business-functions/people-and-organizational-performance/our-insights/why-diversity-matters

04-06 https://library.gv.com/sprint-week-monday-4bf0606b5c81

04-07 https://thoughtbot.com/product-design-sprint/guide/understand/critical-path

04-08 https://racichart.org/the-raci-model/

04-09 https://hired.com/blog/employers/interview-pre-brief-meeting/

04-10 https://www.paradigmiq.com/2020/05/27/how-to-create-more-inclusive-remote-hiring-and-onboarding-processes/

04-11 https://twitter.com/CapitalFactory/status/763205950505353216

04-12 https://twitter.com/SamKap/status/763208168654508033

04-13 https://www.divinc.org/

04-14 http://www.textio.com

04-15 https://bumble.com/jobs

04-16 https://projectinclude.org/hiring#write-inclusive-job-descriptions

04-17 https://github.com/selfdefined/devsofcolour/tree/prod

04-18 https://peopleofcraft.com/

04-19 https://open-learning-resources.myshopify.com/

04-20 https://medium.com/design-ibm/ibm-design-bootcamp-9dda5ed5fb62

04-21 https://www.theguardian.com/lifeandstyle/2019/feb/23/truth-world-built-for-men-car-crashes

04-22 https://medium.com/design-ibm/ibm-design-bootcamp-9dda5ed5fb62

04-23 https://twitter.com/SamKap/status/1366117671893426177

04-24 https://bipartisanpolicy.org/blog/to-pay-or-not-to-pay-cost-distribution-of-the-u-s-employer-based-sponsorship-model/

04-25 https://abookapart.com/products/design-for-cognitive-bias

04-26 https://www.reuters.com/article/us-amazon-com-jobs-automation-Insight/amazon-scraps-secret-ai-recruiting-tool-that-showed-bias-against-women-idUSKCN1MK08G

04-27 https://www.forbes.com/sites/forbeshumanresourcescouncil/2018/04/03/the-benefits-and-shortcomings-of-blind-hiring-in-the-recruitment-process/?sh=55153dfa38a3

04-28 https://www.eeoc.gov/laws/guidance/job-applicants-and-ada

04-29 https://www.netlify.com/careers/

Chapter 5

05-01 https://fortune.com/longform/fortune-500-black-ceos-business-history/

05-02 https://www.census.gov/quickfacts/fact/table/US/PST045219

05-03 http://www3.weforum.org/docs/WEF_GGGR_2021.pdf

05-04 https://www.aiga.org/design/design-research-insights

05-05 https://www.invisionapp.com/inside-design/hurdles-women-design-industry/#:~:text=According%20to%20the%20report%2C%20female,being%20brave%20enough%20to%20ask

05-06 https://projectinclude.org/compensating_fairly

05-07 https://buffer.com/salaries

05-08 https://www.pewresearch.org/fact-tank/2019/12/16/u-s-lacks-mandated-paid-parental-leave/

05-09 https://www.axios.com/2021/09/02/epa-report-climate-change-marginalized-communities

05-10 https://hbr.org/2018/03/for-women-and-minorities-to-get-ahead-managers-must-assign-work-fairly

05-11 https://actreport.com/wp-content/uploads/2021/11/The-ACT-Report.pdf

05-12 https://www.indiebound.org/book/9781732726208

05-13 https://www.kaporcenter.org/wp-content/uploads/2017/04/KAPOR_Tech-Leavers-17-0428.pdf

05-14 https://coco-net.org/problem-woman-colour-nonprofit-organizations/

Chapter 6

06-01 https://medium.com/@george_aye/decolonizing-aiga-a6cc8fb8692e

06-02 https://medium.com/@amelielamont/im-leaving-aiga-behind-you-should-too-f5fa5548bdf8

06-03 https://twitter.com/acarrolldesign/status/1267723742991585281

06-04 https://twitter.com/acarrolldesign/status/1207352021533233153

06-05 https://medium.com/@juan_kafka/my-resignation-from-the-type-directors-club-1023065b4853

06-06 https://technical.ly/delaware/2018/10/03/we-need-to-talk-about-why-jocelyn-harper-left-girl-develop-it/

06-07 https://an-open-letter-to-gdi-board.com/

06-08 https://blog.prototypr.io/why-im-leaving-aiga-d98248aa09d5

06-09 https://womentalkdesign.com/

06-10 https://medium.com/women-talk-design/conference-organizers-can-and-should-help-move-the-needle-on-diversity-8cb91f48e1da

06-11 https://www.confabevents.com/

06-12 http://www.clarityconf.com/

06-13 https://www.clarityconf.com/diversity-and-inclusion

06-14 https://www.clarityconf.com/inclusion

06-15 https://www.confabevents.com/pitching-a-talk

06-16 https://www.refinery29.com/en-us/2018/03/192591/frances-mc-dormand-oscars-speech-inclusion-rider-explainer

06-17 https://gist.github.com/tatianamac/493ca668ee7f-7c07a5b282f6d9132552

06-18 https://ethanmarcotte.com/wrote/my-questions-for-event-organizers/

06-19 https://www.ashedryden.com/blog/codes-of-conduct-101-faq#coc101why

06-20 https://css-tricks.com/psa-linking-to-a-code-of-conduct-template-is-not-the-same-as-having-a-code-of-conduct/

06-21 https://austindesignweek.org/code-of-conduct

06-22 https://queerdesign.club/code-of-conduct/

06-23 https://medium.com/@hellofisher_22863/holding-conferences-accountable-questions-everyone-should-be-asking-bb310ac1c193

06-24 http://rachelnabors.com/2015/09/01/code-of-conduct/

06-25 https://geekfeminism.fandom.com/wiki/Conference_anti-harassment/Actions

Resources

07-01 http://selfdefined.app

07-02 https://www.theconsciouskid.org/resources

07-03 https://www.youtube.com/c/BlairImani/

07-04 https://msw.usc.edu/mswusc-blog/diversity-workshop-guide-to-discussing-identity-power-and-privilege/

07-05 https://www.harpersbazaar.com/culture/politics/a22717725/what-is-toxic-white-feminism/

07-06 https://www.jstor.org/stable/42853921

07-07 https://www.penguinrandomhouse.com/books/611537/beyond-the-gender-binary-by-alok-vaid-menon-illustrated-by-ashley-lukashevsky/

07-08 https://samkapila.com/inclusion/

07-09 https://the-ard.com/

07-10 https://righttobe.org/

07-11 https://being-antiracist.com/

07-12 https://changecatalyst.co/allyshipreport/

07-13 https://www.resmaa.com/

07-14 https://1619books.com/#books

07-15 https://www.indiebound.org/book/9780393356045

07-16 https://abookapart.com/products/accessibility-for-everyone

07-17 https://abookapart.com/products/design-for-safety

07-18 https://uxplanet.org/diversity-and-design-how-we-can-shape-a-more-inclusive-industry-3b12999962e

07-19 https://medium.com/thoughts-from-designx/spotlight-on-gender-diversity-in-design-17a03feae39d

07-20 https://sfdesignweek.org/how-to-infuse-diversity-and-inclusion-in-design/

07-21 https://designobserver.com/feature/double-or-nothing-can-designers-erase-the-gender-pay-gap/39813

07-22 https://docs.google.com/document/d/1OvkmLDaJh4yWznC9U-PogTcvROLmfQdJllSnmjiG83gk/edit

07-23 https://bipocdesignhistory.com/

07-24 https://www.a11yproject.com/

07-25 https://airbnb.design/anotherlens/

07-26 https://inclusive.microsoft.design/

07-27 https://www.blackdogandleventhal.com/titles/andy-campbell/queer-x-design/9780762467853/

07-28 https://abookapart.com/products/cross-cultural-design

07-29 https://www.penguinrandomhouse.com/books/612188/this-is-what-i-know-about-art-by-kimberly-drew-illustrated-by-ashley-lukashevsky/

07-30 https://larahogan.me/blog/onsite-interview-loop-template/

07-31 https://www.greenhouse.io/recruiting

07-32 https://workable.com

07-33 https://askearn.org/
07-34 https://www.nytimes.com/wirecutter/reviews/best-transcription-services/
07-35 https://projectinclude.org/
07-36 https://jobs.localjobnetwork.com/disability
07-37 https://www.pocitjobs.com/
07-38 https://abilityjobs.com/
07-39 https://blacktechpipeline.com/
07-40 https://www.careercontessa.com/
07-41 https://www.pdxwit.org/blog/2019/03/18/how-to-write-an-inclusive-job-posting
07-42 https://medium.com/@meb_57007/writing-effective-and-inclusive-job-descriptions-ace2a302f30a
07-43 https://business.linkedin.com/talent-solutions/blog/job-descriptions/2018/5-must-dos-for-writing-inclusive-job-descriptions
07-44 http://gender-decoder.katmatfield.com/about
07-45 https://bookshop.org/lists/key-books-for-the-inclusive-leader
07-46 https://askearn.org/page/inclusion-at-work-a-framework-for-building-a-disability-inclusive-organization
07-47 https://hbr.org/2018/04/women-of-color-get-asked-to-do-more-office-housework-heres-how-they-can-say-no
07-48 https://hbr.org/2018/03/for-women-and-minorities-to-get-ahead-managers-must-assign-work-fairly
07-49 https://www.harpercollinsleadership.com/9781400220021/the-empathetic-workplace/
07-50 https://www.linkedin.com/in/lilyzheng308
07-51 https://www.linkedin.com/feed/update/urn:li:activity:6884556821844115456/
07-52 https://www.dropbox.com/s/2wswgfeivd93vn5/inclusive-meetings.pdf?dl=0
07-53 https://info.paradigmiq.com/tips-for-inclusive-effective-remote-work
07-54 https://www.hello-collective.com/
07-55 https://www.thenovacollective.com/
07-56 https://www.paradigmiq.com

INDEX

A

accommodations 78-79
active listening 32-33
allyship 16-22
appropriation 7-8
Aye, George 109

B

Bardlavens, Timothy 13, 110
Beecham, Marie 20
biases
    affinity 61
    confirmation 11
    explicit and implicit 27
blackfishing 7
Brooker, Rebecca 45
Brown, Angela B. 117
Budig, Michelle J. 6

C

Carroll, Antoinette 109
codes of conduct 124-128
code-switching 14
Connar, Vincent 38
Cox, Fanshen 123
Coyier, Chris 125
Crenshaw, Kimberlé 6
Critical Path exercise 63-64

D

Dainkeh, Fatima 16
design clubs 108-112
design history 44-47
Devarajan, Kumari 7
diversity, equity, and inclusion (DEI) 94-96
Donovan, Mason 98
Drew, Kimberly 43
Dryden, Ashe 125
Du Bois, W. E. B. 20, 44, 46
Dweck, Carol 25

E

employee resource groups (ERGs) 94
employment benefits 86-89
equality and equity 18
equitable compensation 82-83
events 116-128

F

Fisher, Steve 127

G

gatekeeping 2, 36, 75, 108
Goodman, Timothy 21, 71
groupthink 39

H

hiring myths 56-60
Holmes, Kat 39
hooks, bell 29, 30, 37, 38, 48
Hudgins, Lauren 39

I

identity 5-9, 51, 55, 57, 59
intent and impact 15
institutionalized racism 109
intersectionality 6
intervention 33-34

J

Jina Anne 118, 121
job postings 66-73

K

Kaepernick, Colin 17
Kaplan, Mark 98
Kare, Susan 45

King, Coretta Scott 2
King Jr., Dr. Martin Luther 2
Klint, Hilma af 45
Kooka, Bobby 44
Kotagal, Kalpana 122

L

Lamont, Amélie 16, 71, 109
language 31-32, 67-68
leadership audits 91-43
learning styles 39-40
Learning Zone Model 26-27
Lovelace, Ada 44

M

Mac, Tatiana 71, 123
Mappo, Nini 13
Marcotte, Ethan 39, 47, 52, 124
marginalization 9-16
McCloskey, Hanna Naima 57
McCord, Patty 58
McIntosh, Peggy 20
Meggs, Philip 42
microaggressions 10-12, 101
Multhaup, Marina 91

N

Nabors, Rachel 128

O

Oluo, Ijeoma 7

P

Popova, Maria 25
pre-briefs 64
privilege 20-23, 101, 102

R

Rao, Umesh 44
recruiting 71-72
Roetter, Alex 57

S

safety 13-15
salaries 82-84
Scher, Paula 45
Slootman, Frank 57
Smith, Stacy L. 123
Socratic method 40
Souza, Omari 53
stereotypes 2, 10, 11, 14, 24, 57
supporting students 53

T

Tatum, Beverly Daniel 24
Thomas, David Dylan 27, 76
Thompson, Wanna 7
transparency 70-71

V

Vignelli, Massimo 44
Villanueva, Juan 109
Voss, John 45
Vygotsky, Lev 27

W

white supremacy 36
whitewashing 7
Williams, Joan C. 91

Z

Zhou, Youyou 31

COLOPHON

The text is set in Lora and Plein, both by Indian Type Foundry. Headlines and cover are set in Tatsuro by Vocal Type Company.

ABOUT THE AUTHOR

Sameera Kapila is a design leader, educator, speaker, and Third Culture Kid. Born in India and raised on the Caribbean island of Curaçao, Sameera's experience—having held roles in agencies, educational institutions, and consultancies ranging from individual contributor to executive leadership, student to educator, and everything in between—influences and powers her to analyze design holistically, identifying gaps and advocating for solutions. She writes and speaks about web and product design, inclusion and equity, tech education, and design research and process. She has written regularly for The Pastry Box Project and net Magazine and has spoken at Creative Mornings, SXSW's HBCU track, Design/Content, and Clarity. She also serves on multiple advisory boards in design organizations and educational institutions.

NOTES

www.ingramcontent.com/pod-product-compliance
Lightning Source LLC
Chambersburg PA
CBHW052031030426
42337CB00027B/4952